Extracellular
Enzymes of
Microorganisms

Extracellular Enzymes of Microorganisms

Edited by

J. Chaloupka
and
V. Krumphanzl

Institute of Microbiology
Czechoslovak Academy of Sciences
Prague, Czechoslovakia

Plenum Press • New York and London

Library of Congress Cataloging in Publication Data

Extracellular enzymes of microorganisms.

  Includes bibliographical references and index.
  1. Extracellular enzymes—Congresses. I. Chaloupka, Jirí. II. Krumphanzl, V. (Válcav)
QR90.E98 1987   576'.11925   87-14102

ISBN-13: 978-1-4684-1276-5   e-ISBN-13: 978-1-4684-1274-1
DOI: 10.1007/978-1-4684-1274-1

Proceeding of an international symposium on Extracellular Enzymes of Microorganisms,
held September 1-5, 1986, at Bechyne Castle, Czechoslovakia

© 1987 Plenum Press, New York
 Softcover reprint of the hardcover 1st edition 1987
A Division of Plenum Publishing Corporation
233 Spring Street, New York, N.Y. 10013

PREFACE

Czechoslovak Society for Microbiology and Institute of Microbiology of
the Czechoslovak Academy of Sciences organized an international symposium
"Extracellular enzymes of microorganisms" in September 1986. The symposium
took place in a small South-Bohemian town Bechyně and this book includes
the main contributions presented at the meeting.

The study of microbial extracellular enzymes is a rapidly developing
field of science, which is important both from practical and theoretical
point of view. On one hand, microbial enzymes are nowadays broadly used in
various branches of industry, medicine and agriculture, on the other hand,
their study contributes substantially to our knowledge of problems of
protein secretion, regulation of protein synthesis as related with growth
and cytodifferentiation and - last but not least - it brings data important
for the elucidation of evolutionary pathways. Microbial enzymology also
represents a bordering area between different scientific disciplines such
as microbiology, biochemistry, genetics, biotechnology and other and
demonstrates that only their integration brings about a substantial
progress in the development of our understanding of biological processes.

The symposium in Bechyně was a small one but the contributions of
scientists from 15 different countries have brought information on recent
approaches and developments in this field. I hope that this meeting was
useful in facilitating closer contacts between specialists from different
scientific areas and also contributed to a better understanding among
people from different parts of the world.

Prof. Vladimír Krumphanzl
Corresponding Member of the Czechoslovak
Academy of Sciences, Director of the
Institute of Microbiology and President
of the Czechoslovak Society for
Microbiology

CONTENTS

PART V
OTHER ENZYMES

PART I
GENERAL ASPECTS OF ENZYME FORMATION

REGULATION OF EXTRACELLULAR ENZYME

SYNTHESIS IN BACILLI

Fergus G. Priest

Department of Brewing and Biological Sciences
Heriot-Watt University, Edinburgh EH1 1HX

INTRODUCTION

Extracellular enzymes have been produced on an industrial scale from micro-organisms since the beginning of this century. For economic reasons it is sensible to maximize enzyme yield by optimization of the fermentation conditions and consequently a wealth of information concerning the effect of the environment on extracellular enzyme synthesis has been accumulated (reviewed by Ingle and Boyer, 1976, Priest, 1977). These environmental effects can be conveniently categorized into (1) induction of enzyme synthesis, (2) catabolite repression and (3) developmental regulation, but until recently very little was known of the molecular mechanisms involved. This was largely because the organisms such as Bacillus amyloliquefaciens and B. licheniformis used for the commercial production of α-amylase and proteases were not amenable to genetic analysis. With the development of recombinant DNA techniques in B. subtilis, it has proved possible to transfer genes for proteases and - amylase from industrial organisms into this host (see Yamane et al., 1984; Henner et al., 1985 for reviews) and to analyze the structure and expression of these DNA sequences in detail. In this paper, I shall review the current understanding of the physiology of extracellular enzyme synthesis in Bacillus, in the light of these molecular studies.

INDUCTION OF ENZYME SYNTHESIS

Much has been written about the role of induction in the regulation of extracellular enzyme synthesis and it has often been a controversial issue (reviewed by Ingle and Erickson, 1978). This is largely because the induction ratio (induced level of synthesis as a function of the basal, non-induced level of synthesis) varies considerably between different systems. Extracellular enzymes seldom display the thousandfold induction ratio seen for the archetypal inducible enzyme, β-galactosidase in Escherichia coli. Indeed, some representative data for amylase synthesis in various bacteria are shown in Table 1. The enzyme from B. licheniformis is essentially constitutive, The slight variation in activity seen with different growth substrates probably reflects different growth rates and the effect of catabolite repression (Priest and Thirunavukkarasu, 1985). The α-amylases of B. amyloliquefaciens and B. subtilis are similarly constitutive. However, amylase synthesis by Streptomyces limosus is induced by maltose and starch (Table 1). This seems to be typical of amylase synthesis in streptomycetes since it has also been observed in S. hygroscopicus.

Induction of extracellular enzymes at the physiological level has been described by Tsuyumu (1979). Four stages have been suggested and largely confirmed for polygalacturonate lyase (PL) synthesis in Erwinia carotovora. It is suggested that a basal, constitutive level of enzyme is continually excreted. When substrate is present in the environment, it is degraded by the enzyme with the release, and assimilation of the low molecular weight products. It is these products that effect induction of further enzyme synthesis. But it is wasteful to the producer-organism, and beneficial to its competitors should large amounts of assimilable nutrient accumulate in the environment. Thus in E. carotovora, and perhaps in other organisms, excess product leads to inhibition of extracellular enzyme synthesis by catabolite repression. In this way the organism is able to regulate the supply of assimilable nutrient from an external source.

At the molecular level, PL synthesis by Erwinia crysanthemi has been extensively studied. The action of pectate and oligogalacturonate lyases on polygalacturonic acid leads to the formation of galacturonate and 5-keto-4-deoxyuronate (DKI). The catabolism of these two sugars leads (DKI) to the formation of 2-keto-3-deoxygluconate (KDG) (Condemine et al., 1984; Figure 1). KDG is then phosphorylated and cleaved to form pyruvate and triose phosphate. The use of mutants blocked at various points in the degradative pathway has indicated that KDG and 2,5-diketo-3-deoxygluconate (DKII) are the true inducers of PL synthesis (Condemine et al., 1986).

Recently, some regulatory mutants affecting PL synthesis in Er. crysanthemi have been isolated (Hugouvieux-Cotte-Pattat et al., 1986). In constitutive mutants, designated gpiR, PL synthesis was no longer induced by derivatives of pectin. It is suggested that gpiR, is a regulatory gene and the fact that its inactivation led to constitutive synthesis indicates

Table 1. Effect of Carbon Sources on the Synthesis of α-amylase by Bacillus licheniformis and Streptomyces limosus

| Carbon source | B. licheniformis[1] | |
	Specific enzyme activity	Induction ratio
Arabinose	2.4	1.0
Xylose	2.5	1.0
Maltotriose	3.5	1.5
Maltose	3.9	1.6
Lactose	7.8	3.2

| Carbon source | S. limosus[2] | |
	Enzyme activity	Induction ratio
Lactose	3.5	1.0
Glucose	3.0	0.9
Starch	80.0	23.0
Malt extract	100.0	29.0
Maltose	110.0	31.0

[1]Data from Priest and Thirunavukkarasu (1985)
[2]Data from Fairbairn et al. (1985)

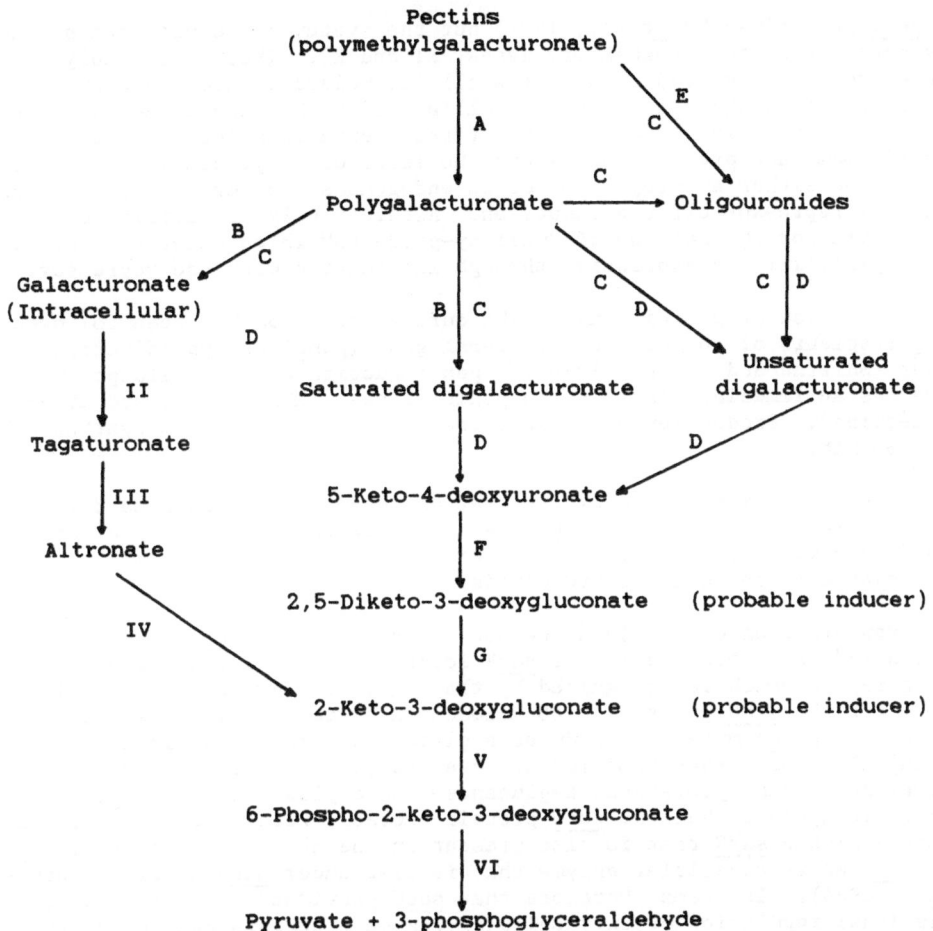

Pectins
(polymethylgalacturonate)

A, pectin methylesterase; B, polygalacturonase; C and E, polygalacturonate and pectin transeliminase; D, oligouronide transeliminase;
F, 5-keto-4-deoxyuronate isomerase; G, 2-keto-3-deoxygluconate oxidoreductase; II, uronate isomerase; III, mannonate oxidoreductase;
IV, altronate hydrolase; V, 2-keto-3-deoxygluconate kinase; VI, 2-
keto-3-deoxy-6-phosphogluconate aldolase.

Fig. 1.

it might code for a repressor. Interestingly, the gpiR mutants also lost
the mechanism that restricts PL synthesis to the stationary phase of the
growth cycle and enzymes were secreted throughout growth (see below). This
suggests that the gpiR product is responsible for both induction of PL
synthesis and its temporal regulation. A second group of constitutive
mutants had similar phenotypes to gpiR but were assigned to kdgR, a locus
that regulates transport of galacturonate into the cell and enzymes V and
VI in Figure 1. The function of kdgR in PL synthesis is however unclear.
Finally, a group of catabolite repression resistant mutants that still
showed inducible synthesis of PL were isolated (Hugouvieux-Cotte-Pattat et
al., 1986).

Inducible extracellular enzymes in Gram positive bacteria have not
been analyzed in the same detail. Penicillinase synthesis in B.
licheniformis is controlled by a trans acting regulatory protein encoded by

the penI gene (Imanaka et al., 1981) but the system is complicated by the presence of two other regulatory genes, R1 and R2. These presumably relate to the problems involved in an extracellular molecule inducing enzyme synthesis within the cell. Indeed Collins (1979) has suggested two models for penicillinase induction. In the first, penicillin interacts with a specific membrane protein and alters the level of a cytoplasmic effector. This may be either a corepressor or an endogenous inducer and by reacting with penI repressor effects induction. Alternatively, penicillin may be responsible for the release of "wall by-products" which enter the cell and induce penicillinase expression through interaction with the repressor.

Modulation of penicillinase mRNA during induction has been followed by using fragments of the cloned structural gene (penP) as hybridization probes (Salerno and Lampen, 1986). Upon induction with cephalosporin C there was an immediate increase in penicillinase mRNA that peaked after 1 h and declined. Production of penicillinase lagged behind mRNA synthesis by about 30 min.

A second Gram positive system that has been studied in some detail is levansucrase synthesis in B. subtilis. This extracellular enzyme catalyses transfructorylation from sucrose to various acceptors which results in levan synthesis and sucrose hydrolysis.

Transcription of the levansucrase structural gene, sacB, is induced by sucrose and is controlled by the sacR locus. This is a cis-acting non-coding region which is recognized by the regulatory protein and is also the target for the sacU protein, a 46 kilodalton protein that, when over-produced in sacU^h mutants, produces a pleiotropic phenotype including loss of flagella, poor transformation and overproduction of levansucrase, neutral and serine proteases, β-glucanase and amylase (Aymerich et al., 1986). It appears that the sacU protein probably recognizes a pallindromic sequence within sacR that is also present in the β-glucanase gene from B. subtilis, an extracellular enzyme that is also under sacU control (Murphy et al., 1984). It seems therefore that sacU provides a complicated transcriptional regulation of extracellular enzyme synthesis but, since the pallindromic sequences appear to be absent from the amylase and protease genes of B. subtilis the function and physiological role of this system remains unclear.

CATABOLITE REPRESSION

Repression of peripheral enzyme synthesis by a rapidly metabolized carbon source such as glucose is an almost universal regulatory phenomenon in micro-organisms. It is a process by which a bacterium may integrate the regulation of its catabolic machinery particularly with regard to energy metabolism. For example, in its normal habitat E. coli is presented with a variety of carbon sources and it would be extremely wasteful if each sugar was to induce synthesis simultaneously of the relevant catabolic enzymes. It seems that to alleviate this problem, promotors for catabolic operons in E. coli form a hierarchy with respect to their affinity for RNA polymerase; those for rapidly-metabolized carbon sources being the strongest and those for poor carbon sources the weakest. Moreover, the weaker the promoter, the higher the requirement for cyclic AMP and its receptor protein CRP for transcription to be initiated. Thus, powerful promoters are recognized by RNA polymerase in the virtual absence of cAMP/CRP and the operons transcribed, if induced. Promoters for poorer carbon sources which require cAMP/CRP for transcription can then be progressively switched on as the growth rate slows and the intracellular cAMP concentration rises until, the poorest carbon sources, which sustains the slowest growth rate (and hence the cAMP concentration is maximal), is metabolized. This somewhat simplistic description of catabolite repression in E. coli largely accounts for a sophisticated integration of carbon metabolism. The delicate balance of promoter strength is emphasized by the lac operon in which a change of just two base pairs in the -10 or Pribnow box sequence (from TATGTT to the canonical sequence TATAAT) alleviates the requirement for cAMP/CRP.

With regard to extracellular enzymes, the PLs of Erwinia species are subject to catabolite repression (Tsuyumu, 1979). Repression is effected by both "good" carbon sources such as glucose but also by unsaturated digalacturonic acids. This "self" catabolite repression by products of the enzyme's activity was incorporated by Tsuyumu into his scheme for enzyme induction (see above) as a means of regulating the activity of an enzyme outside the immediate environ of the cell. In Erwinia, like E. coli, cAMP seems to be largely, but perhaps not entirely responsible for the observed phenomena. Indeed, addition of cAMP to cultures relieves glucose repression of PL synthesis.

Catabolite repression of extracellular enzyme synthesis in Bacillus is also commonly observed. The most thoroughly studied systems involve amylase synthesis in B. subtilis and B. licheniformis and there is some debate about the involvement of catabolite repression in B. amyloliquefaciens (see Coleman, this volume; Ingle and Boyer, 1976). It seems likely that glucose repression of amylase synthesis in this bacterium is less severe than in other bacilli, but early work using B. subtilis B20 (later re-classified as B. amyloliquefaciens B20; (Priest, 1981) showed that glucose does inhibit amylase synthesis in this organism, probably at the transcriptional level (Priest, 1975).

The effect of glucose on amylase synthesis in B. licheniformis NCIB 6346 and a mutant (RM10) selected as catabolite repression resistant and able to form a large zone of starch hydrolysis on the starch-glucose agar plate is shown in Figure 2. In the parental organism, amylase synthesis occurs only when the glucose has been virtually exhausted from the medium (it should be remembered that in this organism, unlike B. subtilis, amylase synthesis accompanies exponential growth (Thirunavukkarasu and Priest, 1980). In the RM10 mutant however, glucose uptake appears to be impaired and amylase synthesis accompanies growth in much the same way as in the wild type strain growing with maltose as carbon source. That this might be a glucose uptake mutant was supported by the fact that it is susceptible to catabolite repression of amylase synthesis by glycerol and mannitol (unpublished observation).

Little is known of the molecular basis of catabolite repression in bacilli although it has long been established that cAMP is not involved (see Priest, 1977 for review). However, some progress is being made through the isolation of mutants of B. subtilis resistant to catabolite repression of amylase synthesis (Nicholson and Chambliss, 1985; 1986). Unlike the mutant of B. licheniformis described above, these mutations (gra-5 and gra-10) are altered specifically in the synthesis of α-amylase and secrete enzyme in the presence of glucose. The mutations have been mapped to the amyR locus which is thought to be a promoter region of amyE, the structural gene. Nicholson and Chambliss (1986) have placed the gra-10 mutation on a plasmid contiguous with cat-86, a promoterless gene for chloramphenicol acetyltransferase. In so doing they constructed a strain in which the chromosomal amyE gene was transcribed from a wild-type amyR locus and the plasmid-borne cat-86 gene was transcribed from the gra-10 allele of amyR. In this strain, amylase synthesis was sensitive to glucose repression but CAT synthesis was unaffected by glucose thus confirming the cis-dominant nature of gra-10 (Nicholson and Chambliss, 1986).

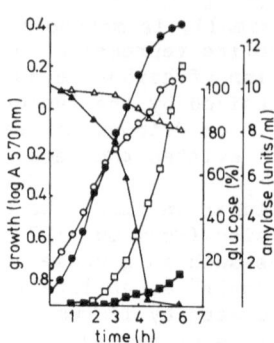

Fig. 2 Growth, O, α-amylase synthesis, □, and glucose consumption △ for Bacillus licheniformis NCIB 6346 (closed symbols) and the catabolite derepressed mutant RMIO (open symbols). Medium was minimal salts (Thirunavukkarascu and Priest, 1984) containing glucose (0.2%) and yeast extract (0.05%).

A particularly important finding related to the gra-10 mutation is that it distinguishes temporal regulation of amylase synthesis in B. subtilis from catabolite repression. An attractive hypothesis for the repression of amylase synthesis during exponential growth and its derepression early in the stationary phase was proposed by Schaeffer (1969) in which, during rapid exponential growth, catabolite repression inhibited amylase synthesis and, as the culture entered stationary phase so catabolite repression was relieved and amylase derepressed. However, in the gra-10 mutant, although amylase synthesis is no longer repressed by glucose, the enzyme remains repressed during exponential growth. This indicates that distinct regulatory mechanisms are responsible for glucose repression and temporal regulation of amylase synthesis. It is interesting to note that mutants of Er. crysanthemi in which pectate lyase synthesis is insensitive to catabolite repression also retain temporal regulation and the enzyme is repressed during exponential growth (see above). Again this distinguishes these two modes of regulation. Moreover, it explains how amylase synthesis in B. licheniformis can accompany exponential growth and yet respond to catabolite repression in the presence of glucose (Thirunavukkarasu and Priest, 1980). If a common mechanism were involved, it would be expected that amylase synthesis in this organism should show temporal regulation.

TEMPORAL REGULATION OF EXTRACELLULAR ENZYME SYNTHESIS

The synthesis of many extracellular enzymes, of which the amylases of B. subtilis and B. amyloliquefaciens are examples is repressed during exponential growth and derepressed as the culture enters stationary phase (see Priest, 1977 for review). Thus extracellular enzyme synthesis coincides with the onset of sporulation, a factor that led to much speculation about the involvement of these enzymes in the sporulation process. The arguments were clarified by Dancer and Mandelstam (1975) who suggested that stationary phase events could be assigned to three categories (i) the primary sequence of dependent events which, if interrupted will prevent sporulation (ii) side events that are initiated by events in (i) but not directly involved in sporulation and (iii) phenotypic changes that are derepressed by those cultural conditions that derepress sporulation.

Mutants of B. subtilis deficient in α-amylase and neutral protease synthesis sporulate normally thus removing these enzymes from category (i). However, serine protease deficient mutants that retained the ability to sporulate proved remarkably difficult to isolate and led to the suggestion that this enzyme may be directly involved in sporulation (category i). However, the recent cloning of the gene for this enzyme followed by in vitro mutagenesis and reinsertion into the chromosome proved that strains that do not synthesize this enzyme can sporulate normally (Stahl and Ferrari, 1984). Thus, no extracellular enzymes are directly involved in sporulation and they appear to have a catabolic function only.

Most sporulation mutants blocked at an early stage (stage zero) of the process (Spo0 mutants) synthesize reduced levels or no serine protease (Ferrari et al., 1986). In particular, spoA mutations completely block serine protease synthesis while other spo0 mutations are less severe and spo0J mutations have virtually no effect. However, amylase and neutral protease are synthesized normally by several spo0 mutants (in particular spo0b and spo0c). This suggests that serine protease is more intimately involved in the sporulation process than the other two enzymes and may be categorized in class (ii) above while amylase and neutral protease may be assigned to class (iii). Should this be the case differences in the regulation of these enzymes might be expected.

RNA polymerase is a heterogeneous enzyme in B. subtilis. The core enzyme can combine with a variety of sigma factors that recognize different promoter sequences (Table 2, Losick and Pero, 1981). These sigma factors are named according to their molecular weights and recently the main vegetative factor, sigma^{55} has been reclassified as sigma^{43} (Doi et al., 1985). It is thought that progressive replacement of these sigma factors and thus changes in transcriptional specificity, is at least party responsible for the process of sporulation.

Genes for several extracellular enzymes of B. subtilis and related organisms have been cloned and sequenced (Table 3). Most of these genes have −35 and −10 sequences similar to the canonical sequence for sigma^{43} and are thus probably transcribed in vivo by the major vegetative form of RNA polymerase. Promoters for serine protease from B. subtilis and probably B. amyloliquefaciens on the other hand resemble the recognition sequence of sigma^{37}, a minor vegetative sigma factor that is also responsible for transcription of the spoVG gene (Moran et al., 1981). The sigma^{37} enzyme appears to increase in activity during the early stationary phase (Moran et al., 1981) which coincides with the synthesis of serine protease. This therefore could explain the more intimate involvement of serine protease with sporulation and the time course of its appearance. However, serine protease expression may be yet more complex since when the promoter is placed as a β-galactosidase fusion in a multi-copy plasmid, the kinetics of synthesis are consistent with the presence of a titratable repressor (Ferrari et al., 1986). However, it should be noted that despite greatly increased β-galactosidase activity from this fusion, enzyme did not appear until after the end of exponential growth.

Table 2. RNA Polymerase of Bacillus subtilis

Sigma factor	Stage of growth	Consensus sequence		
		−35	gap	−10
43 (55)	Major vegetative form	TTGACA	17/18	TATAAT
37	Minor components in	AG--TT	13-16	GG-ATT-TT
32	vegetative cells	AAATC	14-15	TA-TG-TT-TA
28		CTAAA	16	CCGATAT
29	Sporulation	TT-AAA	14-17	CATAAT

Table 3. Promoter Sequences for some Extracellular Enzyme Genes from
Bacilli

Organism	Enzyme	Promoter sequence[1]			Sigma factor	Reference
		−35	gap	−10		
B. amylolique-faciens	Amylase	TTGTTA	17	TATAAT	43	Lehtovaara et al., 1984
	S. protease[3]	GGTCTA	16	TACTAT	37	Wells et al., 1983
B. licheni-formis	Amylase	TTGTTA	17	TACAAC	37	Gray et al., 1986
B. stearo-thermophilus	Amylase	TTGAAA	18	TATAAT	43	Gray et al., 1986 Nakajima et al., 1985
B. substilis	Amylase	TTGATA	17	TAAAAT	43	Yamazaki et al., 1983
	N. protease[2]	TTGAGT	17	TAATAT	43	Henner et al., 1985
	S. protease	AGTCTT	16	TGAATT	37	Wong et al., 1984
	β−glucanase	TTGACC	18	AATCAT	43	Murphy et al., 1984
	Levansucrase	TTGCAA	17	TAGAAT	43	Steimetz et al., 1985
Consensus		TTGACA	17/18	TATAAT	43	Moran et al., 1982

[1]Many of these sequences are presumptive recognition sites
[2]May also contain a sigma[37] sequence
[3]S. protease, serine protease; N. protease, neutral protease

Finally some mechanism must exist for the repression of synthesis of
amylase, neutral protease and related enzymes during exponential growth but
at present the details are unknown.

CONCLUSIONS

Much has been learnt of the regulation of extracellular enzyme
synthesis in bacilli during the past 10 years but still the molecular and
genetic mechanisms remain unclear. Pleiotropic regulatory genes such as
sacU (Aubert et al., 1985), sacQ (Young et al., 1986) and prtR (Nagami and
Tanaka, 1986) have been cloned and sequenced. The control of extracellular
enzyme gene expression is undoubtedly complex but this combined genetic/
biochemical approach is proving to be increasingly effective at unravelling
the details involved.

REFERENCES

Aubert, E., Klier, A. and Rapaport, G., 1985, Cloning and expression in
    Escherichia coli of the regulatory sacU gene from Bacillus subtilis,
    J. Bacteriol., 161:1182.

Aymerich, S., Gonzy-Tréboul, G. and Steinmetz, M., 1986, 5'-Noncoding
region of sacR is the target of all identified regulation affecting
the levansucrase gene in Bacillus subtilis, J. Bacteriol., 166:993.

Collins, J.F., 1979, The Bacillus licheniformis β-lactamase system, in:
"Beta lactamases", J.M.T. Hamilton Miller and J.T. Smith, eds.,
Academic Press, London, pp. 351.

Condemine, G., Hugouvieux-Cotte-Pattat, N. and Robert-Baudy, Y., 1984, An
enzyme in the pectinolytic pathway of Erwinia crysanthemi: 2-keto-3-
deoxygluconate oxidoreductase, J. Gen. Microbiol., 130:2839.

Condemine, G., Hugouvieux-Cotte-Pattat, N. and Robert-Baudy, J., 1986,
Isolation of Erwinia crysanthemi kduD mutants altered in pectin
degradation, J. Bacteriol., 165:937.

Dancer, B.N. and Mandelstam, J., 1975, Criteria for categorizing early
biochemical events during sporulation of Bacillus subtilis, J.
Bacteriol., 121:411.

Doi, R.H., Gitt, M., Wang, L.F., Price, C.W. and Kawamura, F., 1985, Major
sigma factor, sigma-43 of Bacillus subtilis RNA polymerase and
interacting spo0 products are implicated in catabolite control of
sporulation, in: "Molecular Biology of Microbial Differentiation,"
J.A. Hoch and P. Setlow, eds., American Society for Microbiology,
Washington D.C., pp. 157.

Fairbairn, D.A., Priest, F.G. and Stark, J.R., 1986, Extracellular amylase
synthesis by Streptomyces limosus, Enzyme Microb. Technol., 8:89.

Ferrari, E., Howard, S.M.M. and Hoch, J.A., 1986, Effect of stage 0
sporulation mutations on subtilisin expression, J. Bacteriol.,
166:173.

Henner, D.J., Young, M., Band, L. and Wells, J.A., 1985, Expression of
cloned protease genes in Bacillus subtilis, in: "Molecular Biology
of Microbial Differentiation", J.A. Hoch and P. Setlow, eds.,
American Society for Microbiology, Washington D.C. pp. 95.

Hugouvieux-Cotte-Pattat, N., Reverchon, S., Condemine, G. and Robert-Baudy,
J. 1986, Regulatory mutations affecting the synthesis of pectate
lyase in Erwinia crysanthemi, J. Gen. Microbiol., 132:2099.

Imanaka, T., Tanaka, T., Tsunekawa, T. and Aiba, S., 1981, Cloning of the
genes for penicillinase, penP and PenI of Bacillus licheniformis and
their expression in Escherichia coli, Bacillus subtilis and Bacillus
licheniformis, J. Bacteriol., 147:776.

Ingle, M.B. and Boyer, E.W., 1976, Production of industrial enzymes by
Bacillus species, in: "Microbiology 1976", D. Schlessinger, ed.,
American Society for Microbiology, Washington D.C., pp. 420.

Ingle, M.B. and Erickson, R.J., 1978, Bacterial α-amylases, Adv. Appl.
Microbiol., 24:257.

Lehtovaara, P., Vlmanen, I. and Palva, I., 1984, In vivo transcription
initiation and termination sites of an α-amylase gene from Bacillus
amyloliquefaciens cloned in Bacillus subtilis, Gene, 30:11.

Losick, R. and Pero, J., 1981, Cascades of sigma factors, Cell, 25:582.

Moran, C.P., Jr., Lang, N., Banner, C.D.B., Haldewang, W.G. and Losick, R.,
1981, Promoter for a developmentally regulated gene in Bacillus
subtilis, Cell, 25:783.

Moran, C.P., Jr., Lang, N., LeGrice, S.E.J., Lee., G., Stephens, M.,
Sonenshein, A.L., Pero, J. and Losick, R., 1982, Nucleotide
sequences that signal the initiation of transcription and
translation in Bacillus subtilis, Mol. Gen. Genet., 186:339.

Murphy, N., McConnell, D.J. and Cantwell, B.A., 1984, The DNA sequence of
the gene and genetic control sites for the excreted B. subtilis
enzyme β-glucanase, Nucl. A. Res., 12:5355.

Nagami, Y. and Tanaka, T., 1986, Molecular cloning and nucleotide sequence
of a DNA fragment from Bacillus natto that enhances production of
extracellular proteases and levansucrase in Bacillus subtilis, J.
Bacteriol., 166:20

Nakajima, R., Imanaka, T. and Aiba, S., 1985, Nucleotide sequence of the Bacillus stearothermophilus α-amylase gene, J. Bacteriol., 163:401.

Nicholson, W.L. and Chambliss, G.H., 1985, Isolation and characterization of a cis-acting mutation conferring catabolite repression resistance to α-amylase synthesis in Bacillus subtilis, J. Bacteriol., 161:875.

Nicholson, W.L. and Chambliss, G.H., 1986, Molecular cloning of cis-acting regulatory alleles of the Bacillus subtilis amyR region by using gene conversion transformation, J. Bacteriol., 165:663.

Priest, F.G., 1975, Effect of glucose and cyclic nucleotides on the transcription of α-amylase mRNA in Bacillus subtilis, Biochem. Biophys. Res. Commun., 63:606.

Priest, F.G., 1977, Extracellular enzyme synthesis in the genus Bacillus, Bacteriol. Rev., 41:711.

Priest, F.G., 1981, DNA homology in the genus Bacillus, in: "The Aerobic, Endospore-forming Bacteria, Classification and Identification", R.C.W. Berkeley and M. Goodfellow, eds., Academic Press, London, pp. 33.

Priest, F.G. and Thirunavukkarasu, M., 1985, Regulation of α-amylase and α-glucosidase synthesis in batch and chemostat cultures of Bacillus licheniformis, J. Appl. Bacteriol., 58:381.

Salerno, A.J. and Lampen, J.O., 1986, Transcriptional analysis of β-lactamase regulation in Bacillus licheniformis, J. Bacteriol., 166:769.

Schaeffer, P., 1969, Sporulation and the production of antibiotics, exoenzymes and exotoxins, Bacteriol. Rev., 33:48.

Stahl, M. and Ferrari, E., 1984, Replacement of the Bacillus subtilis subtilisin structural gene with an in vitro derived deletion mutation J. Bacteriol., 158:411

Steinmetz, M. LeCoc, D., Aymerich, S., Gonzy Tréboul, G. and Gay, P., 1985, The DNA sequence of the gene for the secreted Bacillus subtilis enzyme levansucrase and its genetic control sites, Mol. Gen. Genet., 200:220.

Stephens, M.A., Ortlepp, S.A., Ollington, J.F. and McConnell, D.J., 1984, Nucleotide sequence of the 5' region of the Bacillus licheniformis α-amylase gene: comparison with the B. amyloliquefaciens gene, J. Bacteriol., 158:369.

Thirunavukkarasu, M. and Priest, F.G., 1980, Regulation of α-amylase synthesis in Bacillus licheniformis NCIB 6346, FEMS Microbiol. Letts., 7:315.

Tsuyumu, S. 1979, "Self catabolite repression" of pectate lyase in Erwinia carotovora, J. Bacteriol., 137:1305.

Wells, J.A., Ferrari, E., Henner, D.J., Estell, D.A. and Chen, E.Y., 1983, Cloning, sequencing and secretion of Bacillus amyloliquefaciens subtilisin in Bacillus subtilis, Nucl. A. Res., 11:7911.

Wong, S.-L., Price, C.W., Goldfarb, D.S. and Doi, R.H., 1984, The subtilisin E gene of Bacillus subtilis is transcribed from a sigma[37] promoter in vivo, Proc. Natl. Acad. Sci. USA, 81:1184.

Yamane, K., Otozai, K., Ohmura, K., Nakayama, A., Yamazaki, H., Yamasaki, M. and Tamura, G., 1984, Secretion vector of Bacillus subtilis constructed from the Bacillus subtilis α-amylase promoter and signal peptide coding region, in: "Genetics and Biotechnology of Bacilli", A.T. Ganasan and J.A. Hoch, eds., Academic Press, London and Orlando, pp. 181.

Yamazaki, H., Ohmura, K., Nakayama, A., Takeichi, Y., Otozai, K., Yamasaki, M., Tamura, G. and Yamane, K., 1983, Amylase genes (amyR2 and amyE) from an α-amylase-hyperproducing Bacillus subtilis strain: molecular cloning and nucleotide sequences, J. Bacteriol., 156:327.

A COMPARISON OF THE CHARACTERISTICS OF EXTRACELLULAR
PROTEIN SECRETION BY A GRAM-POSITIVE AND A GRAM-
NEGATIVE BACTERIUM

G. Coleman, B. Abbas-Ali, J. Sutherland, L. Fyfe
and A. Finley

Department of Biochemistry, Nottingham University
Medical School, Queen's Medical Centre, Clifton Boulevard
Nottingham, NG7 2UH, UK

INTRODUCTION

The background to our present knowledge of the extracellular enzymes
of bacteria was established in a review by Pollock (1962) whose own
interest was in the inducible formation of penicillinase by Bacillus
cereus. At this same time, we developed an interest in a different system
in which extracellular enzyme formation appeared not to be inducible and in
which there was a massive increase in production after the end of the
exponential phase of the growth cycle. Thus, Coleman and Grant (1966)
described the characteristics of alpha-amylase formation by an organism of
the genus Bacillus, obtained as a high alpha-amylase-producing strain of
Bacillus subtilis, in the presence of different carbon sources. During
these early studies, problems were created, inadvertently, and perpetuated
due to confusion over nomenclature. The position was clarified by Welker
and Campbell (1967) who showed that within the genus Bacillus there were
two characteristic patterns of extracellular enzyme formation. Alpha-
amylase was formed, for example, at a very low level throughout the growth
cycle, in one case, and, in the other, a low level of alpha-amylase was
achieved during the exponential phase which increased, massively, in the
post-exponential phase of the growth cycle. A more detailed examination
showed that organisms which produced these two characteristic patters were
distinctly different. The organism known as B. subtilis produced exoenzyme
at a low level throughout the growth cycle whilst the organism which
produced extracellular protein in greatly increased amounts after the end
of exponential growth was shown to be genetically distinct from B. subtilis
and was named Bacillus amyloliquefaciens. Welker and Campbell (1967)
published these findings nearly twenty years ago but in spite of this
attempts are still made to relate data from the two species resulting in
misconceptions and misinterpretations. We have not studied B. subtilis in
any detail other than to confirm the characteristics claimed by Welker and
Campbell (1967) for our strain of B. amyloliquefaciens and B. subtilis 168
(Brown and Coleman, 1975).

Evidence is available which demonstrated that alpha-amylase formation
in B. subtilis is regulated by catabolite repression (Priest, 1975). Thus,
in an organism in which extracellular protein represents a relatively minor
amount of total bacterial protein output, the addition of glucose to the
medium caused a dramatic reduction in exoenzyme formation.

13

Mandelstam (1961, 1962) suggested conditions which were most conducive to the development of catabolite repression. Thus, the addition of glucose to a culture with a high bacterial density, in which growth was no longer taking place, favored the rapid generation of a high concentration of repressing metabolites. Application of these conditions to cultures of B. amyloliquefaciens had no significant effect on extracellular enzyme formation (Coleman, 1967).

More detailed studies on B. amyloliquefaciens showed that, in this system, extracellular protein accounted for a large proportion of the organism's total protein output (Stormonth and Coleman, 1974). Under favorable conditions the maximum rate of extracellular protein formation was equal to the maximum rate of cell protein synthesis. This observation drew attention to the fact that if exoprotein formation was regulated by catabolite repression then, during exponential growth, if the repressing catabolites were removed, it would not be possible to achieve the maximum rate of extracellular enzyme production without considerable reduction in the growth rate.

These considerations and results obtained from measurement of changes in macromolecules and their precursor pools throughout the growth cycle of B. amyloliquefaciens (Stormonth and Coleman, 1974) were used, together with the most reputable data available at the time, to develop the "competition" model of regulation based on the earlier proposal of Coleman (1967). Thus, Coleman et al. (1975) suggested that the rate of total exoprotein formation was coupled to growth by an inverse relationship. Two separate effects were considered to be involved, one of which was superimposable on the other. The first was thought to result from an increase in enzyme concentration on "switching off" non-translatable RNA synthesis and the second an increase in substrate concentration as a result of non-translatable RNA "turnover". Depending on the nutritional environment and the precise conditions of growth it was suggested that these effects could vary in their relative importance in contributing towards increased exoprotein formation in the post-exponential phase of the growth cycle.

The "competition" model provided an explanation of the characteristics of secretion of those extracellular enzymes which are formed at the maximum rate after the end of exponential growth and which account for a significant part of the bacterium's total protein output. A mathematical model of exoprotein production in bacteria has been proposed, recently, which is consistent with the "competition" mechanism (Coleman and Fowler, 1984). It should be emphasized that the ideas implicit in such a mechanism may have been overlooked by those interested in only a single extracellular enzyme rather that the summation of all the individual extracellular proteins secreted by an organism. However, within its framework the opportunity for individual components to respond to other specific regulatory devices is not excluded.

In the foregoing presentation, two regulatory mechanisms were considered both of which might produce the same pattern of exoprotein secretion, that is, a low rate during the exponential phase and a much higher rate afterwards. It is suggested that where exoprotein formation represents a minor fraction of total bacterial protein production catabolite repression may be important whereas in cases where it accounts for a major portion "competition" may be of greater importance. This hypothesis can be examined under the test conditions described by Mandelstam (1961, 1962).

Against this background, interest was developed in the Gram-positive, high alpha-toxin-secreting Staphylococcus aureus (Wood 46) and the Gram-negative, protease-secreting Aeromonas salmonicida (Unilever 2864) both of which produce a large number of extracellular proteins. Such character-istics render these organisms suitable as systems in which to carry out a comparative study of the relationship of growth and total exoprotein syn-thesis to the regulation of formation of specific extracellular products.

METHODS

Organisms

Staphylococcus aureus (Wood 46) (NCTC 7121) and Aeromonas salmonicida (Unilever 2864) were used.

Growth conditions.

Both organisms were grown in TSB medium, consisting of Tryptone Soya Broth (Oxoid Ltd.; 3%, w/v) supplemented with inorganic ions and vitamins as described by Abbas-Ali and Coleman (1977). Batches of medium (50ml), contained in 250 ml conical flasks, were inoculated with bacteria from Tryptone Soya Agar (Oxoid Ltd.) slopes by means of a platinum loop. The cultures were incubated in a "Gyratory" incubator-shaker (model G25, New Brunswick Scientific Co., New Brunswick, N.J., U.S.A.) at 37° C in the case of S. aureus and 25° C for A. salmonicida.

Washed bacterial cell suspension experiments

In order to examine the effect of glucose on extracellular protein formation bacteria from 16 h cultures were harvested by centrifugation at 6,500 x g for 3 min at room temperature, washed once by resuspension in fresh TSB medium, at 25° C, for A. salmonicida, or 37° C, for S. aureus, and finally resuspended in fresh TSB medium, with or without the addition of 1% (w/v) glucose, to the original culture volume. The washed cell suspensions were incubated at 25° C or 37° C, with shaking, and samples were taken for assay at hourly intervals for up to 6 h (Coleman, 1983).

Bacterial density determination

The method of Stormonth and Coleman (1972) was employed.

Extracellular protein estimation.

The protein content of culture supernatant fractions was determined by the method of Sedmak and Grossberg (1977) as described by Coleman et al. (1978).

SDS-polyacrylamide gel electrophoresis

The procedure described by Laemmli (1970) was employed. Samples were run in parallel with a prepared mixture of molecular weight marker proteins (MW-SDS-70L; Sigma Chemical Co. Ltd.).

RESULTS

Comparison of the characteristics of growth and exoprotein secretion

Both organisms grew in TSB medium at their optimal temperatures of 25° C, for A. salmonicida, and 37° C, for S. aureus, to reach the stationary phase by 18 h when bacterial densities of 2.4 and 3.6 mg dry wt/ml, respect-ively, were achieved. The end of exponential growth occurred after 12 h at a bacterial density of 1.0 mg dry wt/ml in the case of <u>A. salmonicida</u> compared with 8 h and 1.9 mg dry wt/ml with <u>S. aureus</u>.

In both cases, the rate of exoprotein formation increased after the end of exponential growth, the change in the case of <u>A. salmonicida</u> being much more pronounced. This is best illustrated by comparing the differential rates of exoprotein formation (Monod <u>et al.</u>, 1952) in the exponential and post-exponential phases of the growth cycle. In the case of <u>A. salmonicida</u> there was a tenfold increase compared with the fourfold increase observed in the <u>S. aureus</u> system. After 24 h the protein concentrations in the <u>A. salmonicida</u> and <u>S. aureus</u> culture supernatant fractions were 0.10 and 0.48 mg/ml, respectively.

It was of considerable interest to examine the progress of increase of exoprotein to these levels in comparison with the increase in cell protein. In Figure 1 the time courses of cell and extracellular protein formation are expressed as percentage values of the maximum levels of total protein produced in the two cultures. It can be seen that in <u>S. aureus</u> the maximum rate of exoprotein formation achieved was equal to the maximum rate of cell protein formation and after 24 h exoprotein accounted for 30% of the total protein output in the culture (Abbas–Ali and Coleman, 1977). By comparison, <u>A. salmonicida</u> achieved a maximum rate of cell protein formation 25 times greater than the rate of exoprotein formation which after 24 h accounted for only 6% of the total bacterial protein production.

<u>Comparison of the effect of glucose on extracellular protein formation by washed bacterial suspensions</u>

Post-exponential phase cultures of the two organisms were centrifuged and the bacterial pellets washed and resuspended in fresh TSB medium, with or without 1% glucose, to the original density. The resulting suspensions were incubated at their optimal temperatures of 25° or 37°C and samples taken for assay at hourly intervals.

In the absence of glucose, in both cases, the bacterial density increased to a plateau during the period of the experiment, at a level approximately 50% higher that the zerotime value. The addition of glucose produced a further 20 to 25% increase in bacterial density in each case.

The characteristics of extracellular protein formation in the two systems were quite different. Thus, in the case of <u>S. aureus</u>, as previously shown by Coleman (1983), a biphasic differential rate of exoprotein formation was observed which, initially, was lower in the presence of the 1% glucose addition. Subsequently, similar high values were achieved irrespective of the presence of the 1% glucose supplement, as shown in Table 1. In the case of <u>A. salmonicida</u>, in the absence of added glucose their was a biphasic differential rate of exoprotein formation, however, when glucose was added a lower differential rate was achieved which diminished to zero during the course of the incubation (Table 1).

The extracellular proteins produced in the various washed bacterial suspensions were visualized by SDS-polyacrylamide gel electrophoresis. The resulting patterns are shown in Figure 2. The extracellular proteins produced by <u>S. aureus</u> are shown in track 5, without added glucose, and a track 6, with 1% added glucose, both tracks show the production of a large number of extracellular proteins with many common components. The most noticeable difference was in the loss of a major band at 33K in the presence of glucose, this corresponds in a position to alpha-toxin which was shown to be repressed by the presence of glucose in an earlier study (Coleman, 1983). The corresponding results for <u>A. salmonicida</u> are visualized in tracks 3 (no added glucose) and 4 (plus 1% glucose). It can be

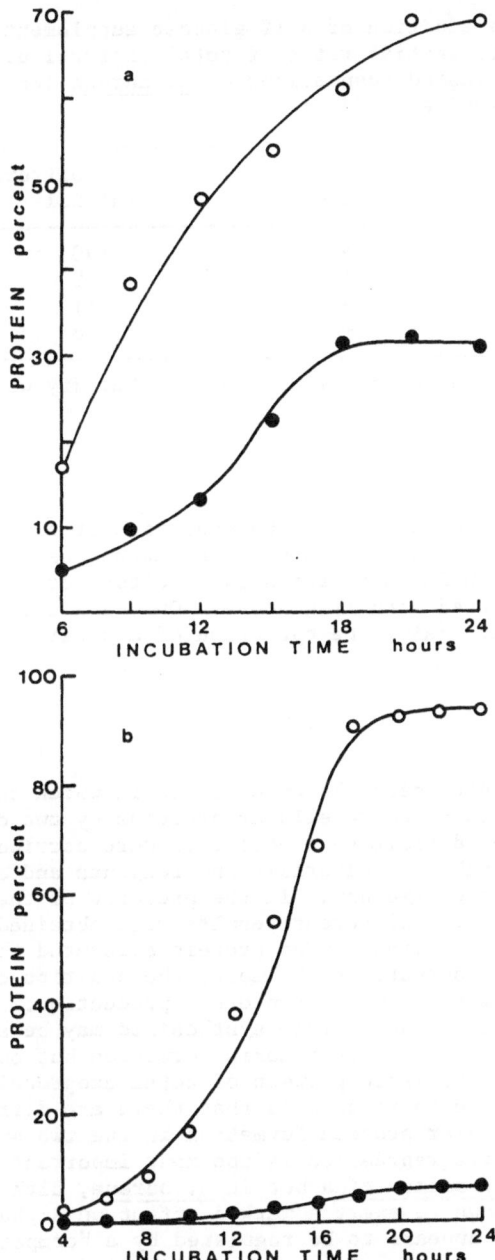

Fig. 1. Distribution of protein between cellular (o) and extracellular (●) phases during the growth of (a) S. aureus (Wood 46) and (b) A. salmonicida (Unilever 2864). The results are expressed as percentages of the total protein produced by the culture after 24 h growth.

seen that a large number of exoproteins were secreted in the absence of added glucose in a washed bacterial suspension, in fact, the pattern is almost identical to that produced in a 24 h culture of the organism in TSB medium shown in track 1, with the characteristic major protease band at 70K. However, as shown in track 4, few proteins were formed in the

17

Table 1. Effect of the addition of a 1% glucose supplement on the initial and final differential rates of total extracellular protein formation by washed suspensions of S. aureus (Wood 46) and A. salmonicida (Unilever 2864)

Organism	Glucose added	Differential rate*	
		initial	final
S. aureus	-	190	560
	+	70	520
A. salmonicida	-	11	44
	+	8	0

* Expressed in terms of µg exoprotein formed per mg dry wt increased in bacterial mass.

presence of added glucose suggesting the operation of a very marked glucose effect or catabolite repression. This, observation was supported by the pattern in track 2 which is that from a 24 h culture of the organism grown in TSB medium supplemented with 1% glucose. Only a single faint band can be seen at 70K showing, again, the operation of a powerful catabolite repression.

DISCUSSION

The results presented describe experiments in which the effect of glucose on the formation of extracellular proteins by two different organisms was examined. The definitive experiments were carried out at the same time using the same batches of materials and reagents and all the assays were done together on the same day. In the presence of these precautions, as shown in Figure 2, quite different results were obtained. Thus, with A. salmonicida, in which extracellular protein accounted for 6% of the total bacterial protein output, at the most, the addition of glucose caused a virtually complete repression of exoprotein production. In the case of S. aureus, in which 30% of the protein synthesized may be secreted, there was a very marked reduction in alpha-toxin formation but only a minor modulation of the characteristic pattern of total exoprotein synthesis. The obvious conclusion to be reached is that there are differences in the regulation of extracellular protein formation in the two systems. It is suggested that catabolite repression is the most important exoprotein regulatory device in A. salmonicida but in S. aureus, although catabolite repression has been shown to exert a marked effect on alpha-toxin formation, total exoprotein appears to be regulated by a "competition" mechanism.

In its simplest terms, catabolite repression results from the presence of catabolites in the bacterial culture medium, this produces a reduction in the affinity of RNA polymerase for a promoter causing a reduction in the differential rate of enzyme formation (Wanner et al., 1978). Such a repression may occur without change in growth characteristics or even without change in bacterial density.

On the other hand, a reduction in extracellular protein formation accountable to a mechanism based on the idea of "competition" (Coleman et al., 1975) may be caused by the presence of catabolites in the medium only when their direct effect is on growth of the organism. This, in turn, will produce an indirect effect at the level of exoprotein gene expression.

18

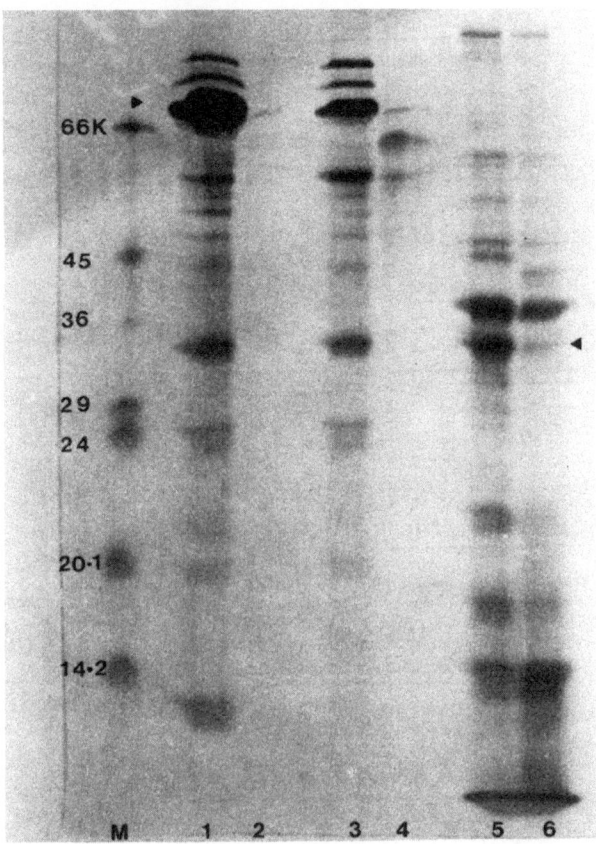

Fig. 2.   SDS-polyacrylamide gel electrophoresis patterns of the
          extracellular proteins produced by S. aureus (Wood 46) and A.
          salmonicida (Unilever 2864) in the presence and absence of a 1%
          glucose supplement.
          Tracks 1 and 2, 24 h culture of A. salmonicida without glucose and
          with 1% glucose; tracks 3 and 4, A. salmonicida washed cell
          suspension without and with 1% glucose; tracks 5 and 6, S. aureus
          washed cell suspension without and with 1% glucose.
          Molecular weight marker proteins (MW-SDS-70L, Sigma) are separated
          in track M.

In other words, no change in growth rate - no change in characteristics of
extracellular protein formation.

     Thus, the difference between the two mechanisms, as illustrated in
Figure 3, is that catabolite repression is a direct result of the presence
or increase of a catabolite in the affected cell. Control by "competition"
is an indirect effect produced by catabolites which directly influence
growth. This effect on growth is considered to be accompanied by changes
in the concentrations of components of the exoprotein transcriptional
machinery (Stormonth and Coleman, 1974).

     Since the two mechanisms are different there is no reason why they
should not exist side-by-side and operate independently (Coleman, 1983).

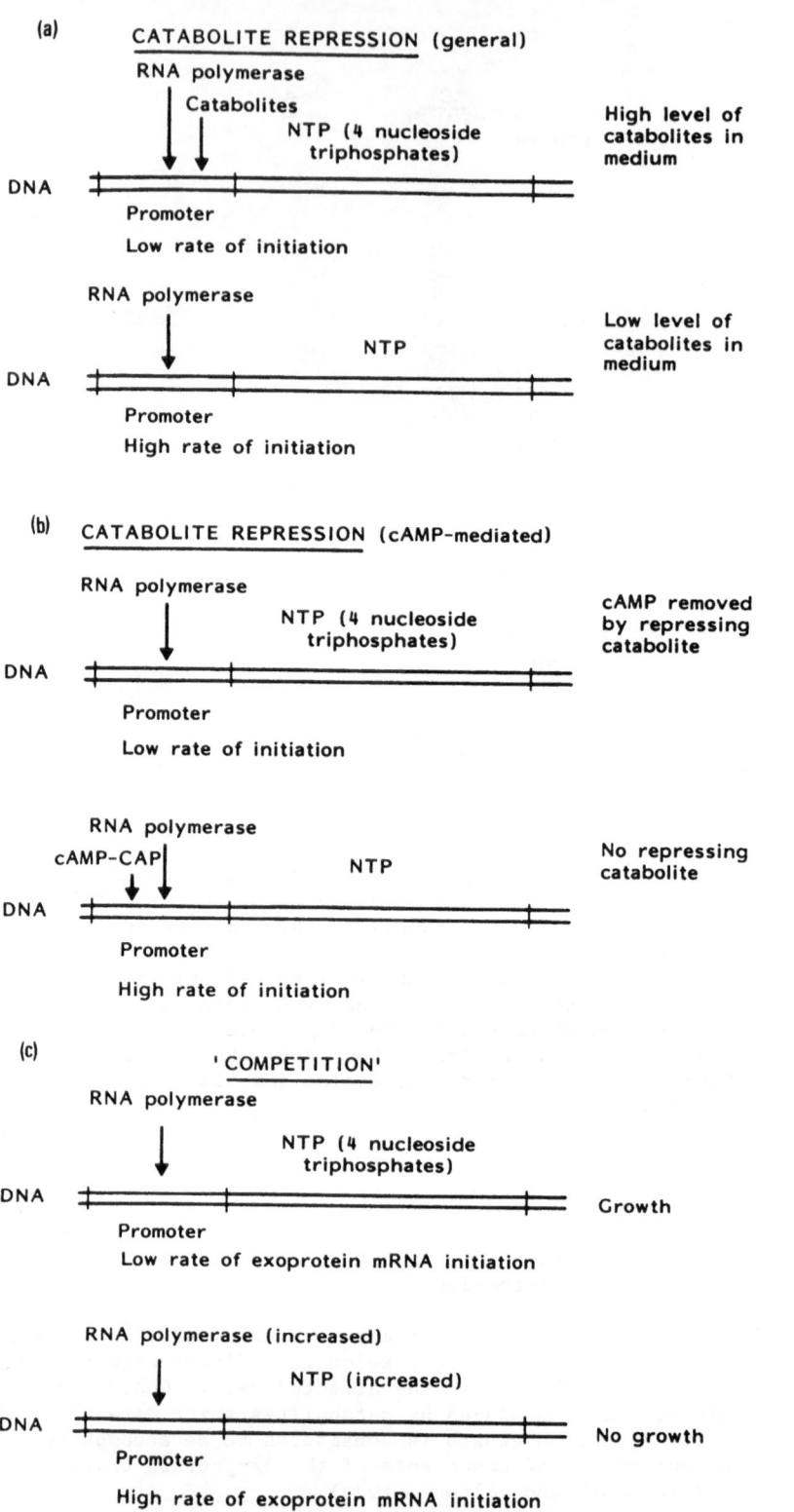

Fig. 3. Comparative schemes relating (a) general and (b) cAMP-mediated catabolite repression with (c) "competition".

Acknowledgments

A.F., L.F. and J.S. wish to thank the Science and Engineering Research Council for Research Studentships.

Figure 1 (a) is published with the permission of the Journal of General Microbiology.

REFERENCES

Abbas-Ali, B., and Coleman, G., 1977, Characteristics of extracellular protein secretion by Staphylococcus aureus (Wood 46) and their relationship to the regulation of alpha toxin formation., J. Gen. Microbiol., 99:277.

Brown, S., and Coleman, G., 1975, Messenger ribonucleic acid content of Bacillus amyloliquefaciens throughout its growth cycle compared with Bacillus subtilis 168., J. Mol. Biol., 96:345.

Coleman, G., 1967, Studies on the regulation of extracellular enzyme formation by Bacillus subtilis., J. Gen. Microbiol., 49:421.

Coleman, G., 1983, The effect of glucose on the differential rates of extracellular protein and alpha-toxin formation by Staphylococcus aureus (Wood 46)., Arch. Microbiol., 134:208.

Coleman, G., and Grant, M.A., 1966, Characteristics of alpha-amylase formation by Bacillus subtilis., Nature, London, 211:306.

Coleman, G., Brown, S., and Stormonth, D.A., 1975, A model for the regulation of bacterial extracellular enzyme and toxin biosynthesis., J. Theoret. Biol., 52: 143.

Coleman, G., Jakeman, C.M., and Martin, N., 1978, Patterns of total extracellular protein secretion by a number of clinically isolated strains of Staphylococcus aureus. J. Gen. Microbiol., 107:189.

Coleman, K.D., and Fowler, A.C., 1984, A mathematical model of exoprotein production in bacteria., IMA J. Math. Appl. Med. Biol., 1:77.

Laemmli, U.K., 1970, Cleavage of structural proteins during the assembly of the head of bacteriophage T4., Nature, London, 227:680.

Mandelstam, J., 1961, Induction and repression of beta-galactosidase in non-growing Escherichia coli., Biochem. J., 79:489.

Mandelstam, J., 1962, The repression of constitutive beta-galactosidase in Escherichia coli by glucose and other carbon sources., Biochem. J., 82: 489.

Monod, J., Pappenheimer, A.M., and Cohen-Bazire, G., 1952, La cinétique de la biosynthèse de la beta-galactosidase chez E. coli considérée comme fonction de la croissance., Biochim. Biophys. Acta, 9:648.

Pollock, M.R., 1962, Exoenzymes. in: "The Bacteria", vol. 4, I.C. Gunsalus and R.Y. Stanier, ed., Academic Press Inc., New York.

Priest, F.G., 1975, Effect of glucose and cyclic nucleotides on the transcription of alpha-amylase mRNA in Bacillus subtilis., Biochem. Biophys. Res. Commun., 63:606

Sedmak, J.J., and Grossberg, S.E., 1977, A rapid, sensitive and versatile assay for protein using Coomassie Brilliant Blue G250., Anal. Biochem., 79:544.

Stormonth, D.A., and Coleman, G., 1972, A rapid and convenient method for the determination of cell densities of bacteria which form aggregates., J. Gen. Microbiol., 71:407.

Stormonth, D.A., and Coleman, G., 1974, Cellular changes accompanying the transition from minimal to maximal rate of exoenzyme secretion by Bacillus amyloliquefaciens., J. Appl. Bacteriol., 37:225.

Wanner, B.L., Kodaira, R., and Neidhardt, F.C., 1978, Regulation of the lac operon expression: Reappraisal of the theory of catabolite repression., J. Bacteriol., 136:947.

Welker, N.E., and Campbell, L.L., 1967, Unrelatedness of Bacillus amyloliquefaciens and Bacillus subtilis., J. Bacteriol., 94:1124.

MODELLING OF PHYSIOLOGICAL CONTROL OF PRODUCTION

OF EXOCELLULAR HYDROLYTIC ENZYMES

J. Votruba, J. Pazlarová and J. Chaloupka

Departments of Enzyme and Process Engineering
Institute of Microbiology
Czechoslovak Acad. Sci. Prague 4

INTRODUCTION

The rate of production of several exocellular hydrolytic enzymes (e.g. amylases, lipases and several proteinases) is generally connected with physiological functions coupled with growth. In this paper an attempt has been made to design a mathematical model of an observed kinetic of enzyme production. Figure 1 shows the typical course of growth and enzyme production during batch culture. From the viewpoint of microbial physiology the course of growth and enzyme production can be split into four regions. In region I the lag in growth and enzyme formation is observed. In region II the growth is maximal while the production is delayed and has a transient character. In region III the growth is slower due to exhaustion of the limiting substrate but the rate of enzyme formation is at its maximum. The transient period of growth is short but most of hydrolytic enzymes are produced during that period. When the limiting substrate is exhausted the growth ceases and the rate of enzyme production can be zero or negative due to decay processes. The kinetics of decay in this period of cultivation is usually of first order [1] with a half-time of 5 - 8 hours. Only some extracellular enzymes, whose synthesis proceeds during sporogenesis (e.g. serine proteinase in Bacillus subtilis) are produced by non-growing populations.

In this way we can describe the observed physiology of growth and enzyme production on a level of population. For better understanding of the rate controlling mechanisms it is useful to describe the physiology of enzyme production during the individual cell cycle. Mitchison [2] in his comprehensive book described four different types of synthesis of an individual protein during the cell cycle. (cf. Figure 2) The first case is a stepwise synthesis of an enzyme. During a certain part of the cell cycle the rate of synthesis rises to its maximum and ceases when another part of cell cycle is reached (Figure 2a). The second type depicted in Figure 2b is a peak-like enzyme synthesis. The observed rate of synthesis rises in certain part of cell cycle and reaches its maximum. After this maximum the synthesis ceases and decay of the enzyme takes place. In other two cases (Figure 2c, d) a piecewise linear or exponential synthesis is observed. From the described physiology of individual protein synthesis during the cell cycle according to Mitchison [2] the periodicity in this synthesis based on duration of the cell cycle is an apparent phenomenon. These commonly known experimental findings on enzyme synthesis on population and

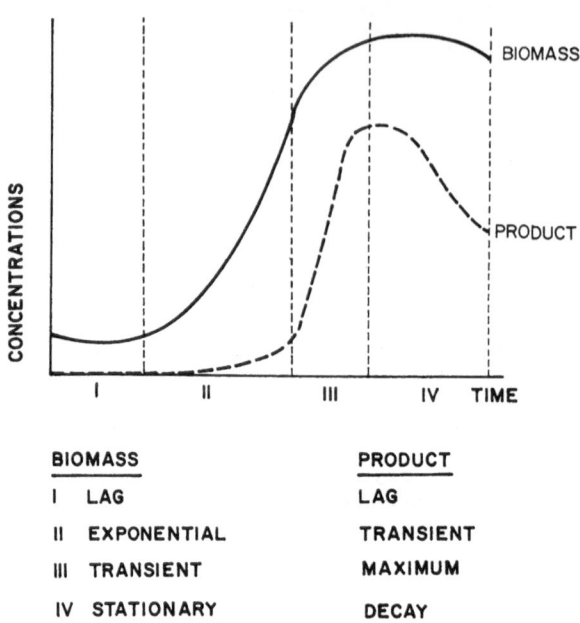

BIOMASS	PRODUCT
I LAG	LAG
II EXPONENTIAL	TRANSIENT
III TRANSIENT	MAXIMUM
IV STATIONARY	DECAY

Fig. 1. A typical course of growth and exocellular enzyme production in a batch culture. The fermentation is split into four regions describing the physiology of growth and product formation.

individual cell cycle level are used in the next part of this paper in the design of a mathematical model describing the observed overall kinetics of the exocellular hydrolytic enzyme production.

Formulation of the Model

Figure 3 shows two cases of fast and slow growth. The period in which DNA replication takes place is constant for both cases [2,3]. The length of the period in which the exocellular hydrolytic enzymes are synthetized is controlled by the environment. The time of "opened doors" for enzyme synthesis is proportional to the certain portion of cell cycle. When rapid growth takes place the effective period for "opened doors" for the enzyme synthesis is shorter and the resulting enzyme synthesis is lower. When slower growth occurs, the effective time period for enzyme synthesis is longer and then the resulting amount of the product is higher in comparison with fast growth. This formal model does not distinguish between a non-specific regulation by the growth rate and a specific one due to a repression of enzyme synthesis caused by some component of a rich medium. If refer back to Figure 1 this model logically fits the observed courses of biomass and product formation in the batch culture. When the observed growth is close to zero (the lag and the stationary phase of growth) the length of the cell cycle is extremely long and an individual cell is usually at the state before the "opened doors" period for enzyme synthesis. According to our model the enzyme synthesis is zero. The case of exponential and transient growth fits our scheme depicted in Figure 3. The mean length of cell cycle in a population can be assumed to be proportional to the reciprocal value of specific growth rate [2,3]. Then likewise the observed specific production rate is proportional to the mean reciprocal value of the "opened doors" period for enzyme synthesis. The ratio between the mean length of cell cycle and the "opened doors" period for enzyme synthesis is then a certain function of the environment. Thus the general

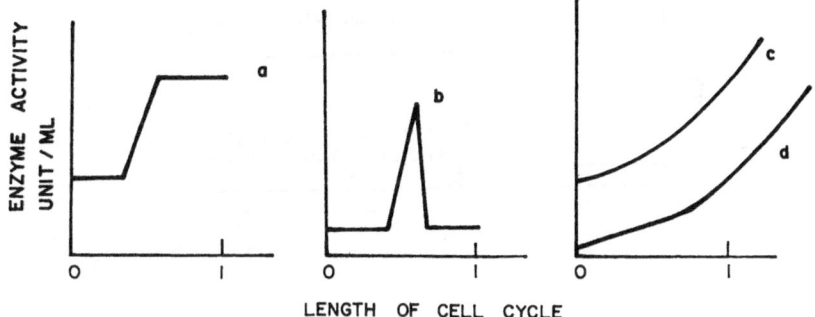

Fig. 2. Types of individual protein synthesis during the cell cycle.

model for physiological control of exocellular enzyme synthesis can be
written in the following way:

$$R_E = f(Environment) \quad \mu X \qquad (1)$$

Where R_E is overall rate of enzyme synthesis, μ is the specific growth
rate, X is biomass concentration and f (Environment) is the function describing the influence of environment on the "opened doors" period.

Identification of the Model

The task of identification of the model (Eq. 1) can be transformed to
the identification of the unknown function f (Environment). According to
equation 1 this function can be evaluated when overall rate of enzyme
synthesis (observed rate + decay rate) is divided by the observed growth
rate for exponential and transient phase of growth. According to the
proposed model this ratio for lag and stationary phase is equal to zero.
The influence of the environment can be characterized by many variables,
e.g. pH, temperature, intensity of external energy fields and/or by chemical composition of cultivation broth. In most experiments the physical
variables are nearly constant and microbial physiology is influenced by the
concentration of the so called limiting substrate. In the review paper of
Koplove and Cooney [4] the effect of different limiting substrates on
enzyme formation during transient growth is well documented and therefore
the concept of the limiting substrate as the controlling environmental
agent may be accepted. Figures 4 and 5 show the calculated values of $R_E/\mu X$
versus the concentration of the growth limiting substrate. The case of
α-amylase production by Bacillus subtilis [5] limited by caseinate is
depicted in Figure 4. Another case of exocellular proteinase production by
Bacillus megaterium [6] is plotted on Figure 5. In both cases the curves
have the same shape which shows an increase in $R_E/\mu X$ ratio when the limiting substrate is exhausted. The difference between the two graphs is in
the value of the basal synthesis rate. This basal synthesis is not influenced by the concentration of the limiting substrate. The basal synthesis
of α-amylase is asymptotically close to zero in contrast to that of the
proteinase.

As was shown previously [5] the course of an unknown function which
describes the effect of a limiting substrate on the enzyme production can
be correlated by a simple exponential function. Taking into account the
basal synthesis not affected by the concentration of the limiting substrate, this unknown function can be described as follows:

$$f(Environment) = R_E/\mu X = K_1 + K_2 \exp(-S/K_S) \qquad (2)$$

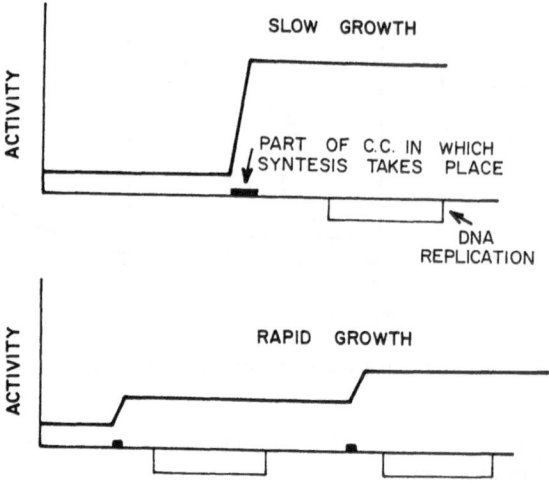

Fig. 3. Schematic model of synthesis of exocellular enzyme during the cell cycle (CC).

Where the first term on the right hand side of Eq. 2 describes the basal synthesis and the second one the synthesis regulated by the substrate concentration. This second term can be considered a substrate repression of the enzyme synthesis. However, looking back on the scheme in Figure 3, this effect may be also related to the mere prolongation of the "opened doors" period during the individual and mean population cell cycle. On comparing Eqs. 1 and 2 the final version of model for physiological control of exocellular enzyme synthesis can be described in the following way:

$$R = [K_1 + K_2 \exp(-S/K_S)] \mu X - K_3 E \tag{3}$$

The observed rate of production R is equal to the formation rate minus the enzyme decay rate, where X is biomass concentration, S is the concentration of the limiting substrate, E is the activity of the enzyme, μ is the specific growth rate and K_1, K_2, K_3, K_S are unknown parameters which have to be estimated from experimental data. With both α-amylase and proteinase production the cultivation was performed on synthetic media where the concentration of limiting substrate were known and measurable. By contrast, in most exocellular enzyme productions complex media with an unknown concentration of unidentified limiting substrates are used. The model can be simplified when the mass balance for the limiting substrate S is applied as follows:

$$dS/dt = -Y_{X/S} dX/dt \tag{4}$$

This equation can be integrated for initial condition describing substrate exhaustion (S=0 at X=X$_{maximal}$). On integration we obtained the relation between the measured biomass concentration X and substrate concentration S:

$$S = (X_{max} - X) / Y_{X/S} \tag{5}$$

Replacing S in Eq. 3 by Eq. 5 we obtain a modified formula for the observed rate of enzyme formation:

$$R = [K_1 + K_2 \exp (-\frac{X_{max} - X}{K_X})] \mu X - K_3 E \tag{6}$$

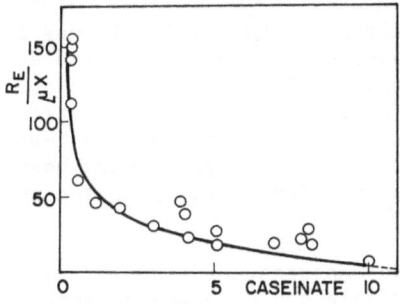

Fig. 4. Influence of environment as a function of the concentration of the growth limiting substrate (caseinate g/l) for α-amylase produced by Bacillus subtilis in batch culture.

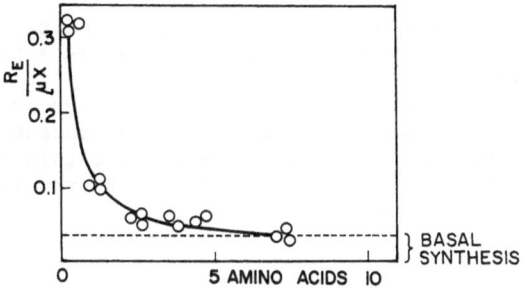

Fig. 5. Influence of environment as a function of the concentration of the growth limiting substrate (aminoacids g/l) for proteinase produced by Bacillus megaterium in batch culture.

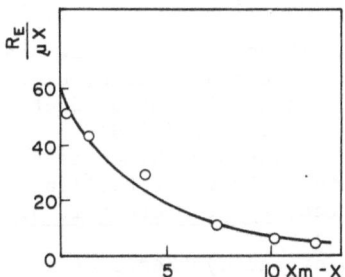

Fig. 6. Influence of environment as a function of $X_{max} - X$ for lipase produced by Aspergilus niger in batch culture.

This transformed relation was applied to the production of exocellular lipase [7]. The values of $R_E/\mu X$ were plotted against $X_{max} - X$ in Figure 6. The course of the function describing the influence of environment on the production of lipase is comparable with the dependences for the limiting substrate in Figures 4 and 5. Equation 6 for enzyme production simplifies the model of physiological control and can be applied in various situations for evaluation and computer simulation of hydrolytic enzyme production.

CONCLUSION

After the identification of the proposed model for physiological control of α-amylase, proteinase and lipase production in a batch culture an attempt was made to generalize the model. It was found that the function describing the influence of environment on the enzyme production exhibits the same shape for α-amylase synthesis in continuous culture [8] and therefore the effect of different age distribution in a population is not important. The same situation was found with β-amylase in a batch culture and for other enzymes such as lysine decarboxylase or V-penicillin amidase. The more general validity of the proposed model is being verified.

This model has its own importance for optimal design of process control. The controlling variable could be the concentration of the growth limiting substrate while the controlled one is the rate of production in the transient phase of growth. As was shown for the case of α-amylase [5] a fed-batch culture seems to be very promising for increasing enzyme production. For this case a proper feeding strategy based on the proposed model for the "opened doors" period of enzyme synthesis leads to a doubling of production in comparison with a classical batch culture.

The fed-batch technique of cultivation is limited by fermenter volume because the volume of fermentation broth increases after each substrate addition. This disadvantage can be avoided by application a cyclic feeding strategy [9-11] which allows one to control the synchrony of individual cell development [9] by the limiting substrate concentration. The preliminary simulation results on a computer show that this type of environmental control of continuous production of exocellular hydrolytic enzymes based on our model is very promising.

REFERENCES

1. R. T. Dean, "Cellular degradative processes", Chapman and Hall, London (1978)
2. J. M. Mitchison, "The biology of cell cycle", University Press, Cambridge (1973)
3. V. N. Ivanov and G.A. Ugotchikov, "Cell cycle and heterogenity of microbial populations", Naukovaya dumka, Kiev (1984)
4. M. H. Koplove and Ch. L. Cooney, Enzyme production during transient growth, Adv. Biochem. Eng., 12:2 (1979)
5. J. Pazlarová, M.A. Baig, J. Votruba, Kinetics of α-amylase production in a batch and fed-batch culture of Bacillus subtilis with caseinate as a nitrogen source, Appl. Microbiol. Biotechnol., 20: 331 (1984)
6. J. Chaloupka, Regulation of the synthesis of extracellular proteinases in bacilli, in "Environmental Regulation of Microbial Metabolism", J.S. Kulaev, E.A. Dawes and D.W. Tempest, eds., Academic Press, London, pp. 287 (1985)
7. N. Pal, S. Das and A.K. Kundu, Influence of culture and nutritional conditions on the production of lipase by submerged culture of Aspergillus niger, J. Ferment. Technol., 56:593 (1978)
8. E. J. Emanuilova and K. Toda, α-amylase production in batch and continuous cultures by Bacillus caldolytic, Appl. Microbiol. Biotechnol., 19:301 (1984)
9. P. S.S. Dawson, Continuous fermentation - and new horizons, Can. J. Chem. Eng., 62:293 (1984)
10. K. Vu-Trong and P.O. Grady, Stimulation of tylosin productivity resulting from cyclic feeding profiles in fed-batch cultures, Biotechnol. Lett., 4:725 (1982)
11. B. Heinritz, G. Rogge, E. Stichel and Th. Bley, Use of biorythmic processes for increasing the efficiency of carbon-compound conversion by microorganisms, Acta Biotechnol. 3:125 (1983)

CLONING, SEQUENCING, AND EXPRESSION IN <u>E. COLI</u>

AND <u>B. SUBTILIS</u> OF A STAPHYLOKINASE GENE

Detlev Behnke*, Dieter Gerlach*, and Regine Kraft**

* Zentralinstitut für Mikrobiologie und experimentelle
Therapie, Akademie der Wissenschaften der DDR
6900 Jena, PSF 73, DDR
**Zentralinstitut für Molekularbiologie
Akademie der Wissenschaften der DDR
1115 Berlin-Buch, Robert-Rössle-Str. 10, DDR

INTRODUCTION

Substances that activate plasminogen have recently attracted considerable attention because of their potential application as drugs in thrombolytic therapy. Staphylokinase (SAK) elaborated by many strains of <u>S.
aureus</u> is one of the plasminogen activating substances of bacterial origin
[1,2]. Unlike the mammalian plasminogen activators urokinase [3] and
tissue plasminogen activator [4] it lacks detectable proteolytic activity
or any other known enzymatic property [5]. Activation of plasminogen by
SAK occurs rather by a stoichiometric association of the two proteins which
then results in a conversion of plasminogen to plasmin. SAK shares these
peculiarities with another bacterial plasminogen activator-streptokinase,
which is already commercially available as a drug [6]. Extensive studies
on SAK have been hampered in the past by difficulties to purify sufficient
amounts of the protein from contaminating staphylococcal toxins.

Production of SAK is frequently associated with a lysogenic state of
the producer strain. Introduction of certain bacteriophages into SAK-
negative <u>S. aureus</u> strains results in lysogenic conversion towards SAK
production [7]. Cloning experiments carried out by Sako et al. [8]
convincingly proved the location of a <u>sak</u> gene on the genome of a
staphylococcal phage.

We report here the cloning of a second <u>sak</u> gene derived from staphylococcal phage 42D which is representative of serogroup F phages and which
has previously been shown to cause lysogenic conversion of staphylococcal
strains to SAK production [7,9]. After the initial cloning steps in <u>E.
coli</u> hosts we introduced the <u>sak</u> 42D gene into the heterologous gram-
positive host <u>B. subtilis</u> in order to obtain both expression and secretion
of this protein.

RESULTS

The Assay System

The assay employed to visualize SAK production was based on the
proteolytic activity of human plasmin [10]. Bacterial clones forming a
plasminogen activator, e.g., SAK, were surrounded by a clear zone of
caseinolysis on agar plates containing both skim milk (1%) and human
plasminogen (10 μg/ml). They failed, however, to exhibit proteolytic
activity on agar plates supplemented with skim milk alone. Differential
testing on these two types of agar was taken as preliminary proof for the
production of SAK (Figure 1).

Cloning of the sak 42D Gene in E. coli

Phage 42D - a standard phage used for lysotyping of S. aureus strains
- served as the source of the sak gene to be cloned. Purified DNA of this
phage (approx. size: 42 kbp) was cleaved with the restriction endonucleases
HpaII, ClaI or EcoRI and the fragments which ranged in size between 0.5 and
10 kbp were cloned into the unique EcoRI or ClaI sites of the standard E.
coli plasmid vector pACYC184 [11]. Among about 500 transformants of E.
coli MM294 tested 11 showed a positive response on SAK assay plates. All
but one of these transformants carried recombinant plasmids which harbored
multiple Ø42D DNA fragments. The only exception was plasmid pDB3 containing a single ClaI fragment of 7 kbp from Ø42D (Figure 2). This particular
Ø42D DNA fragment was also present on all other recombinant plasmids
generated with ClaI. The Ø42D insert of pDB3 contained a single EcoRI and
two EcoRV recognition sites conveniently located for subcloning steps
(Figure 2). Transfer of a 2.5 kbp Ø42D DNA fragment flanked by EcoRI and
EcoRV sites from pDB3 onto pBR322 gave rise to plasmid pDB12 (Figure 2).
This plasmid still conferred upon its host the ability to produce SAK.
Subcloning of a 1.05 kbp RsaI fragment from within the 42D insert of pDB12
into the EcoRV site of pBR322 allowed to further narrow down the location
of the sak 42D gene. The recombinant plasmid pDB17, a detailed restriction
map of which is shown in Figure 3, had suffered a deletion during the
cloning process. This deletion removed essentially all of the pBR322

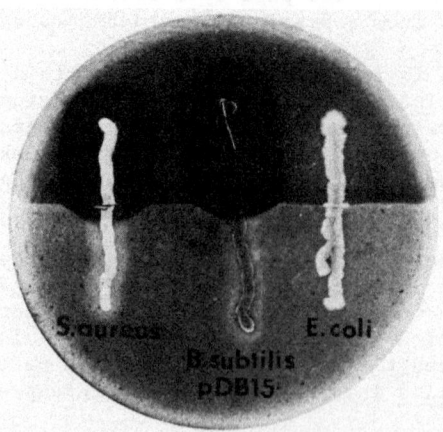

Fig. 1. Assay for SAK activity. The agar plate contained: yeast extract
(0.1%), casamino acids (0.05%), and either skim milk (1%) alone
(lower half) or together with human plasminogen (10 μg/ml; upper
half). Caseinolysis due to activated plasminogen occurs
specifically only on the upper part.

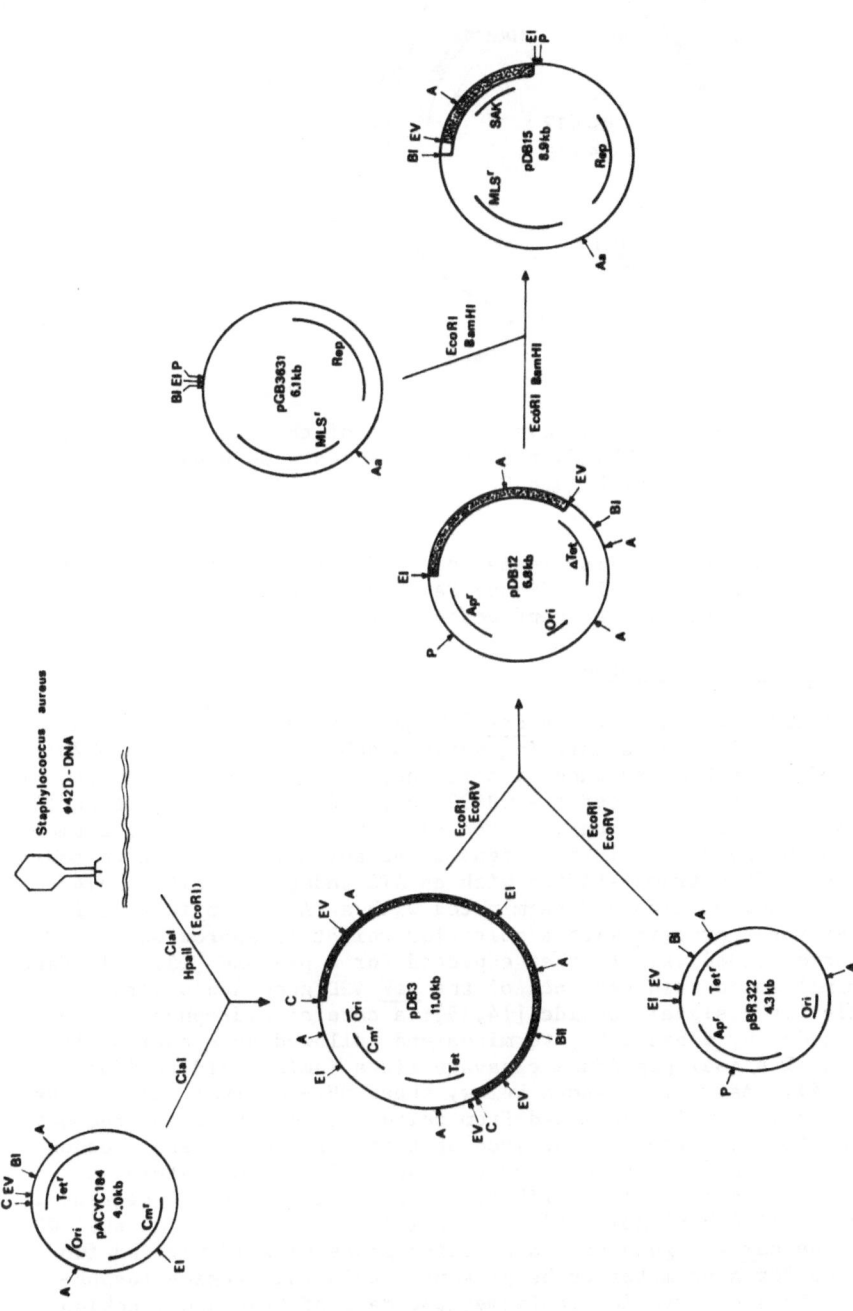

Fig. 2. Cloning scheme for the sak42D gene. Dotted areas represent $\emptyset$42D DNA fragments. Cm^R-chloramphenicol resistance, Tet^R-tetracycline resistance, Ap^R-ampicillin resistance, MLS^R-resistance to macrolides, lincosamids and streptogramin, Ori-origin of replication, Rep-region essential for replication, SAK-staphylokinase, restriction enzyme abbreviations: Aa-AvaI, A-AccI, BI-BamHI, BII-BglII, C-ClaI, EI-EcoRI, EV-EcoRV, P-PstI.

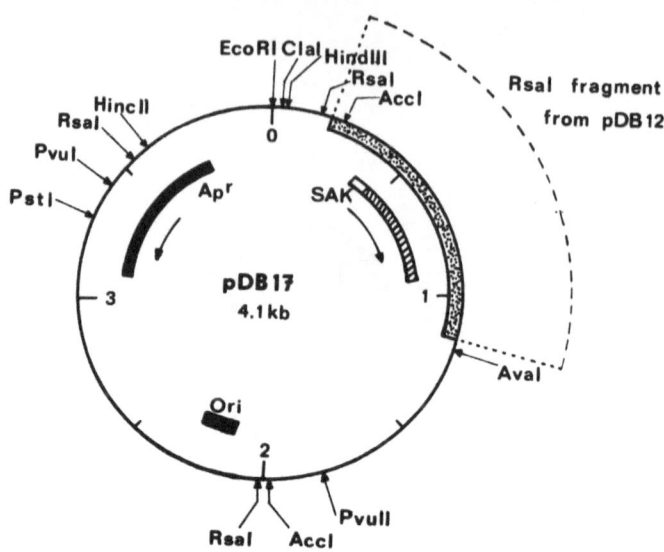

Fig. 3.　Restriction enzyme cleavage site map of the sak-recombinant
plasmid pDB17. The dotted area represents Ø42D-DNA. For
abbreviations see Figure 2.

sequences coding for tetracycline resistance and part of the Ø42D insert.
Nevertheless, the plasmid still mediated SAK production and, thus, had
acquired at least the structural gene coding for SAK.

DNA Sequence of the sak 42D Gene

The nucleotide sequence of the sak 42D gene was determined by dideoxy
sequencing [12] of M13 clones carrying various subfragments of the Ø42D
insert of pDB17. The DNA sequence of this insert is shown in Figure 4. It
comprised 1027 bp which replaced the pBR322 sequences between position 187
and 1412. Analysis of the sequence revealed only one open reading frame
preceded by appropriate translation signals and sufficiently long to code
for SAK. This reading frame started with an ATG codon in position 314,
extended for 163 amino acids and terminated with a TAA codon in position
803. It coded for a protein with a molecular weight of approximately 18.5
kDal which corresponded well to that expected for a pre-SAK [13]. In fact
the NH_2-terminal amino acid sequence of the sak 42D gene had a structure
characteristic for a signal peptide [14,15]: a core of hydrophobic amino
acids was preceded by a basic NH_2-terminus and followed by a region with
several potential signal peptidase cleavage sites (amino acid residues
25-30; Figure 4). As will be shown below, these NH_2-terminal amino acids
were indeed absent from SAK purified from culture supernatants of recombinant Bacillus strains. The DNA sequence upstream of the ATG start codon
was extremely AT-rich (85%) except for the nucleotides immediately
preceding the ATG codon where the ribosome binding sequence (SD-sequence,
binding energy - 14.4 kcal/mol [16]) was positioned. Transfer of the 300
bp preceding the sak 42D gene onto a promoter probe plasmid yielded functional evidence for a promoter to be present within this region (unpublished data). This observation confirmed the data of Sako and Tsuchida
[17] who also located a potential promoter within the corresponding region
of the sak SØC gene by in vitro run-off transcription experiments. The
location of the only potential promoter structure [18] for σ^{43} or σ^{70} RNA
polymerases, respectively, present within the sequence upstream of the sak
42D gene is indicated in Figure 4.

DNA sequence of the Ø24D staphylokinase gene

1   GTATACGCGC   CTGGAACATT   AATATATGTG   TTTGAAATTA   TAGATGGTTG   TTGTCGCATT

TATTGGAACA   ATCATAATGA   GTGGATATGG   CATGAGAGAT   TGATTGTGAA   AGAAGTGTTT

TAATTCTAAG   GTTAAAATGT   TAAATATTTG   TTAATTATTT   TTTAATGTAA   GTTTAGTTTC

                    -35   - 10
TTTTAATATT   TTATTGATTT   TTAATATTTT   CTCAATATAA   AATGAAGTTG   TTGATATTTA

TCATCTTAAA   TAAGGGTGTT   AGCTATAAAA   AGAGATAAAT   AAAAACAAAT   ATATTATATT

SD   1   10
TGGAGGAAGC   GCC ATG CTC AAA AGA AGT TTA TTA TTT TTA ACT GTT TTA TTG TTA TTA
               Met Leu Lys Arg Ser Leu Leu Phe Leu Thr Val Leu Leu Leu Leu

                        20   30
TTC TCA TTT TCT TCA ATT ACT AAT GAG GTA AGT GCA TCA AGT TCA TTC GAC AAA GGA AAA
Phe Ser Phe Ser Ser Ile Thr Asn Glu Val Ser Ala Ser Ser Ser Phe Asp Lys Gly Lys

            40   50
TAT AAA AAA GGC GAT GAC GCG AGT TAT TTT GAA CCA ACA GGC CCG TAT TTG ATG GTA AAT
Tyr Lys Lys Gly Asp Asp Ala Ser Tyr Phe Glu Pro Thr Gly Pro Tyr Leu Met Val Asn

                    60   70
GTG ACT GGA GTT GAT GGT AAA AGA AAT GAA TTG CTA TCC CCT CGT TAT GTC GAG TTT CCT
Val Thr Gly Val Asp Gly Lys Arg Asn Glu Leu Leu Ser Pro Arg Tyr Val Glu Phe Pro

                        80   90
ATT AAA CCT GGG ACT ACA CTT ACA AAA GAA AAA ATT GAA TAC TAT GTC GAA TGG GCA TTA
Ile Lys Pro Gly Thr Thr Leu Thr Lys Glu Lys Ile Glu Tyr Tyr Val Glu Trp Ala Leu

                        100   110
GAT GCG ACA GCA TAT AAA GAG TTT AGA GTA GTT GAA TTA GAT CCA AGC GCA AAG ATC GAA
Asp Ala Thr Ala Tyr Lys Glu Phe Arg Val Val Glu Leu Asp Pro Ser Ala Lys Ile Glu

                        120   130
GTC ACT TAT TAT GAT AAG AAT AAG AAA AAA GAA GAA ACG AAG TCT TTC CCT ATA ACA GAA
Val Thr Tyr Tyr Asp Lys Asn Lys Lys Lys Glu Glu Thr Lys Ser Phe Pro Ile Thr Glu

                    140   150
AAA GGT TTT GTT GTC CCA GAT TTA TCA GAG CAT ATT AAA AAC CCT GGA TTC AAC TTA ATT
Lys Gly Phe Val Val Pro Asp Leu Ser Glu His Ile Lys Asn Pro Gly Phe Asn Leu Ile

                    160   Stop
ACA AAG GTT GTT ATA GAA AAG AAA TAA   AACAAAATAG   TTGTTTATTA   TAGAAAGCAA
Thr Lys Val Val Ile Glu Lys Lys

TGTCTTGCTT   GAATATGTGT   AGTGAAAATT   ATCTTTCATC   AAATTCTCAT   TCATGCACGA

ATGGCTCTTC   CCCACCTAAT   CAGATATTAG   GTGACTTATG   GGGAGAAATC   AGTTAGGATA
                                                                    ↑

AAAAGTGGAT   AATCCTTTTT   TTAGGCAGGT   TCCAGGCA

Fig. 4.   DNA-sequence of the Ø42D DNA fragment present on pDB17. The first
          34 amino acids have been omitted to facilitate comparison with the
          sak SØC sequence determined by Sako and Tsuchida [17]. The
          antisense strand of sak 42D is shown. Numbering above the DNA
          sequence refers to amino acid residues. Direct and inverted
          repeats are indicated by arrows above the sequence. Nucleotides

and amino acids differing between the sak 42D and sak SØC sequences are underlined or boxed. Beyond nucleotide 951 (arrow) there is a complete lack of homology between the two genes. SD-Shine/Dalgarno sequence, -35 and -10 denote the location of the potential promoter structure.

Several direct and inverted repeats of various length were scattered over the whole sequence. Whether these repeat structures have any functional significance for promoter efficiency or transcriptional termination remains to be studied.

The DNA sequences of the sak genes from phage 42D and phage SØC [17] were identical to a large extent. Within the region of homology 11 differences were, however, observed either as single nucleotide exchanges or insertions. Three of these differences occurred within the coding sequence causing replacements of two amino acids and one pheno typically silent mutation (see boxed amino acids in Figure 4). The nucleotide sequences of the cloned fragments from Ø42D and SØC differed completely beyond position 951 of the Ø42D sequence shown in Figure 4.

Expression of the sak 42D Gene in Bacillus subtilis

B. subtilis is known for its ability to secrete a variety of extra-cellular enzymes and its lack of toxin production. We therefore introduced the sak 42D gene into this heterologous gram-positive host in order to obtain secretion and thus facilitate purification of the SAK formed. This subcloning was accomplished by transferring the complete Ø42D insert of pDB12 onto plasmid pGB3631 (Figure 2) - a vector designed in this laboratory for molecular cloning in gram-positive bacteria [19] (unpublished data). The resulting recombinant plasmid pDB15 replicated in B. subtilis and mediated SAK production. The Bacillus strains DB104 [20] and GB500 (B. Adler, personal communication) used as recipients exhibited drastically reduced levels of exoprotease. SAK production and secretion by strains DB104 (pDB15) or GB500 (pDB15) initiated at the early phase of growth and continued through the logarithmic period. The amount of active SAK present in the culture supernatant declined soon after the stationary phase was reached because of residual production of exoproteases by the host strains. A yield of approximately 15-20 µg SAK/ml of culture supernatant was obtained at the end of the exponential phase. This amount corresponded to a release of about 3×10^5 molecules of SAK per B. subtilis cell.

The identity of the plasminogen activator present in culture supernatants of recombinant B. subtilis strains with SAK was demonstrated by an Ouchterlony immunoprecipitation test with monospecific anti-SAK-antiserum raised in rabbits against authentic SAK produced by an S. aureus strain (Figure 5).

Characterization of SAK produced by B. subtilis

SAK was purified from culture supernatants of B. subtilis GB500 (pDB15) by first precipitating the protein with TCA and ammonium sulfate and subsequent separation on CM-Sepharose. A final purification step on Sephacryl 200 yielded a SAK preparation that was virtually free of contaminating proteins as judged from SDS-PAA gel electrophoresis (Figure 6). The two protein fractions present in this preparation both activated plasminogen and were serologically identified as SAK. Chromatofocusing of the purified SAK on a Pharmacia PBE94 column yielded, however, three protein species all of which reacted with anti-SAK-antiserum (Figure 7). The larger minority protein seen on PAA gels (Figure 6) corresponded to peak I of the chromatofocusing profile shown in Figure 7. It had an iso-

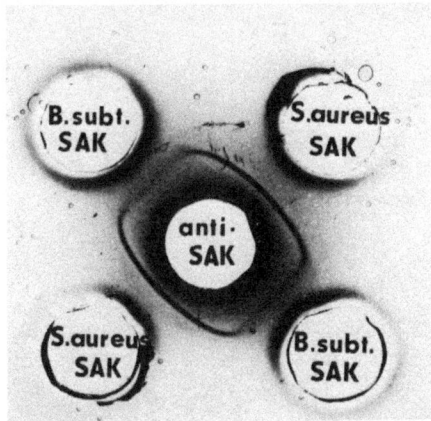

Fig. 5. Ouchterlony immunoprecipitation test of culture supernatants from
the recombinant B. subtilis strain GB500 (pDB15) against
monospecific anti-SAK-antiserum raised in rabbits against
authentic SAK.

electric point (IEP) of 7.6 and was found to be formylated. Most likely it
represented unprocessed pre-SAK. The majority protein present in the SAK
preparation (Figure 6) consisted of two species with IEP's of 6.3 - 6.4
(fraction II) and 5.8 - 5.9 (fraction III, Figure 7). Determination of
their NH_2-terminal amino acid sequence by manual Edman degradation [21]
revealed that they differed in a single residue at their amino terminus.
The additional Lys residue present on fraction II SAK may explain the
higher IEP of this protein.

DISCUSSION

The sak gene of S. aureus phage 42D has been cloned, sequenced and
expressed in E. coli and B. subtilis. The heterologous gram-positive host
proved to be particularly useful since it supported not only the expression
of SAK but also its secretion to the surrounding medium. Efficient
expression and secretion have recently also been reported for other
staphylococcal exoproteins, the genes of which had been cloned and
introduced into Bacillus [22-25]. In contrast to most natural exoproteins
of bacilli which are expressed only after the transition from the
logarithmic to the stationary phase of growth [26] SAK was produced and
secreted already from the onset of logarithmic growth. Logarithmically
growing Bacillus cells therefore must be endowed with all properties
necessary for protein secretion. SAK, like other staphylococcal proteins
may, thus constitute a good model to study protein secretion in bacilli.
Furthermore, the expression and secretion signals of these proteins may
prove suitable for the construction of expression/secretion vectors for
Bacillus since their activity is not linked to certain growth phases. In
fact, we have recently obtained efficient expression in B. subtilis of a
mammalian protein under the control of sak 42D expression signals
(unpublished data).

The sak genes of phages 42D and SØC [17] were identical to a large
extent. The two differences between the amino acid sequences of which at
least one was a non-conservative exchange may reflect a low functional
significance of the protein domains affected. They also explain the
slightly differing IEP's of the two SAKs [27]. Phage 42D SAK secreted by
B. subtilis lacked 38 NH_2-terminal amino acids. This was surprising since
several potential signal peptide processing sites [9,10] were located
further upstream between amino acid positions 25 and 30. Furthermore, Sako

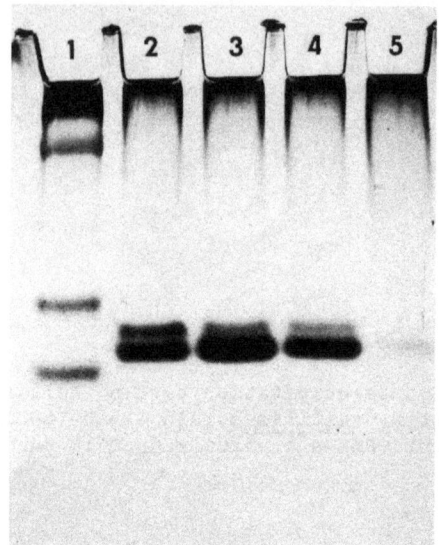

Fig. 6. SDS-PAGE of SAK purified from culture supernatants of <u>B. subtilis</u> GB500 (pDB15). Gel concentration was 10%. Lane 1: molecular weight markers of 67,000; 43,000; 18,000 and 12,000 dal. Lanes 2-4: SAK purified from <u>B. subtilis</u>. Lane 5: SAK purified from <u>S. aureus</u>.

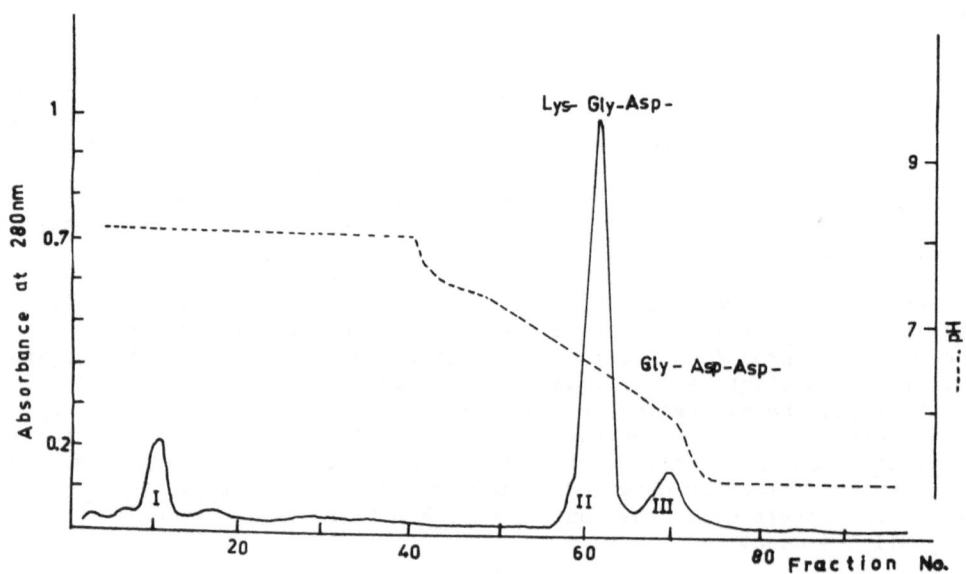

Fig. 7. Chromatofocusing of SAK purified from culture supernatants of <u>B. subtilis</u> GB500 (pDB15). Column: Polybuffer exchanger PBE94, 20 x 1 cm. Sample: 4.6 ml of SAK (4 mg/ml in start buffer). Elution buffer: Polybuffer 96 (30%) + Polybuffer 74 (70%) - CH_3COOH, pH 5.0. Peak I: formylated pre-SAK, IEP 7.6; Peak II: processed SAK NH_2-terminal sequence Lys-Gly-Asp-Asp-Ala-Ser-Tyr-, IEP 6.3 - 6.4; Peak III: processed SAK, NH_2-terminal sequence Gly-Asp-Asp-Ala-Ser-Tyr-, IEP 5.8 - 5.9.

[27] found the major signal processing site of SØC SAK secreted into the periplasm of E. coli cells to be located at the expected position beyond amino acid 27, although a minority fraction of their SAK also initiated further downstream at amino acid 38. Whether this difference in the NH$_2$-terminal part of the two SAKs reflected differences between the protein secretion machinery of B. subtilis and E. coli or whether it was rather due to a postsecretory modification of SAK remains to be resolved. Replacement of SAK sequences beyond amino acid 31 did, however, not impair protein secretion from B. subtilis directed by the sak 42D signal peptide (unpublished data). This observation makes a postsecretory modification the more likely explanation. The heterogeneity of the amino terminus of SAK secreted by Bacillus was not surprising. Similar observations have also been made for natural exoproteins of bacilli [15].

The multitude of extracellular toxins secreted by S. aureus strains has so far been the major obstacle for intensive studies on SAK. The availability of cloned sak genes in more suitable host backgrounds and the efficient expression of this protein will certainly pave the way for more detailed studies which eventually should allow a conclusive evaluation of the potential use of SAK as a drug in thrombolytic therapy.

Acknowledgements

We would like to thank Kai Engelmann and Edda Etzold for dedicated technical assistance.

REFERENCES

1. C. H. Lack, Staphylokinase: An Activator of Plasma Protease, Nature, 161:559-560 (1948).
2. R. Christie and H. Wilson, A Test of Staphylococcal Fibrinolysin, Anst. J. Exp. Biol. Med. Sci., 19:329 (1941).
3. W. Christensen and S. Müllertz, Kinetic Studies of Urokinase Catalysed Conversion on NH$_2$-Terminal Lysin Plasminogen to Plasmin, Biochim. Biophys. Acta, 480:275-281 (1977).
4. M. Rånby, N. Bergsdorf, and T. Nilson, Enzymatic Properties of One- and Two-Chain Forms of Tissue Plasminogen Activator, Throm. Res., 27:175-183 (1982).
5. C. H. Lack and K. L. A. Glanville, Staphylokinase, in: "Methods in Enzymology, Vol. XIX", G. R. Perlman and L. Lorand, eds., Academic Press, New York (1970).
6. F. B. Taylor and R. H. Tomar, Streptokinase, in: "Methods in Enzymology, Vol. XIX", G.P. Perlman and L. Lorand, eds., Academic Press, New York (1970).
7. K. C. Winkler, J. De Waart, C. Grootsen, B. J. M. Zegers, N. F. Tellier and C. D. Vertegt, Lysogenic Conversion of Staphylococci to Loss of β-toxin, J. Gen. Microbiol., 39:321-333 (1965).
8. T. Sako, S. Sawaki, T. Sakurai, S. Ito, Y. Yoshizawa and I. Kondo, Cloning and Expression of the Staphylokinase Gene of Staphylococcus Aureus in Escherichia coli, Mol. Gen. Genet., 190:271-277 (1983).
9. I. Kondo and K. Fujise, Serotype B Staphylococcal Bacteriophage Singly Converting Staphylokinase, Infect. Immun., 18:266-272 (1977).
10. O. Saksela, Radial Caseinolysis in Agarose: A Simple Method for Detection of Plasminogen Activators in the Presence of Inhibitory Substances and Serum, Analyt. Biochem., 111:276-282 (1981).
11. A. C. Y. Chang and S. N. Cohen, Construction and Characterization of Amplifiable Multicopy DNA Cloning Vehicles Derived from the P15A Cryptic Miniplasmid, J. Bacteriol., 134:1141-1156 (1978).
12. F. Sanger, S. Nicklen and A. R. Coulson, DNA-Sequencing with Chain Terminating Inhibitors, Proc. Natl. Acad. Sci. USA, 74:5463-5467 (1977).

13. I. Kondo, S. Itoh and T. Takagi, Purification of Staphylokinase by Affinity Chromatography with Human Plasminogen, in: "Staphylococci and Staphylococcal Infections", Zbl. Bakt. Suppl. 10, J. Jelaszewicz, ed., Gustav Fischer Verlag, Stuttgart (1981).

14. D. Oliver, Protein Secretion in E. coli, Ann. Rev. Microbiol., 39:615-648 (1985).

15. P. S. F. Mezes and J. O. Lampen, Secretion of Proteins by Bacilli, in: "The Molecular Biology of the Bacilli", D. Dubnau, ed., Academic Press, New York City (1985).

16. I. Tinoco, P. N. Borer, B. Dengler, M. D. Levine, O. C. Uhlenbeck, D. M. Crothers and J. Gralla, Improved Estimation of Secondary Structure in Ribonucleic Acids, Nature (New Biol.), 246:40-41 (1973).

17. T. Sako and N. Tsuchida, Nucleotide Sequence of the Staphylokinase Gene from Staphylococcus aureus, Nucl. Acids Res., 11:7679-7693 (1983).

18. M. Rosenberg and D. Court, Regulatory Sequences Involved in the Promotion and Termination of RNA Transcription, Ann. Rev. Genet., 13:319-353 (1979).

19. D. Behnke and M. S. Gilmore, Location of Antibiotic Resistance Determinants, Copy Control and Replication Functions on the Double-Selective Streptococcal Cloning Vector pGB301, Mol. Gen. Genet., 184:115-120 (1981).

20. F. Kawamura and R. H. Doi, Construction of a Bacillus subtilis Double Mutant Deficient in Extracellular Alkaline and Neutral Proteases, J. Bacteriol., 160:442-444 (1984).

21. J. Y. Chang, D. Brauer and B. Wittman-Liebold, Micro-Sequence Analysis of Peptids and Proteins using 4-NN-dimethyl-aminoazobenzene 4'-isothiocyanate/phenylisothiocyanate Double Coupling Method, FEBS Letters, 93:205-214 (1983).

22. S. Kovacevic, L. E. Veal, H. M. Hsiung and J. R. Miller, Secretion of Staphylococcal Nuclease by Bacillus subtilis, J. Bacteriol., 162:521-528 (1985).

23. N. Fairweather, S. Kennedy, T. J. Foster, M. Kehoe and G. Dougan, Expression of a Cloned Staphylococcus aureus α-hemolysin Determinant in Bacillus subtilis and Staphylococcus aureus, Infect. Immun., 41:1112-1117 (1983).

24. S. R. Fahnestock, Ch. W. Saunders, M. S. Guyer, S. Löfdahl, B. Guss, M. Uhlen and M. Lindberg, Expression of the Staphylococcal Protein A Gene in Bacillus subtilis by Integration of the Intact Gene into the B. subtilis Genome, J. Bacteriol., 165:1011-1014 (1986).

25. Ch. Y. Lee and J. J. Iandolo, Mechanism of Bacteriophage Conversion of Lipase Activity in Staphylococcus aureus, J. Bacteriol., 164:288-293 (1985).

26. F. G. Priest, Extracellular Enzyme Synthesis in the Genus Bacillus, Bacteriol. Rev., 41:711-753 (1977).

27. T. Sako, Overproduction of Staphylokinase in Escherichia coli and its Characterization, Eur. J. Biochem., 149:557-563 (1985).

EFFECT OF CERULENIN ON BACILLUS SUBTILIS 168

Svetla Baykousheva, Nadja Cherepova
and Konstantina Ilieva

Institute of Microbiology, Bulgarian Academy
of Sciences, 1113 Sofia, Bulgaria

INTRODUCTION

The antibiotic cerulenin, a specific inhibitor of fatty acid synthesis, prevents the formation of extracellular but not of membrane proteins in some bacteria [1-3]. It has been suggested that this effect is a result of a physico-chemical interaction between the amphiphilic molecule of the agent and the membrane rather than of its interference with lipid synthesis. We used specific biochemical and ultracytochemical methods to study the location and the activity of membrane-bound ATPase in whole cells of Bacillus subtilis grown with cerulenin and in isolated membranes from this microorganism treated in vitro with the drug. The enzyme could serve not only as a sensitive probe for detecting changes induced by membrane-active agents, but in view of some recent findings [4] it seems to be somehow involved in protein translocation across membranes.

METHODS

Growth conditions for cultivation of B. subtilis 168, isolation of membranes and ultracytochemical methods have been described in our previous publications [5,6]. The ATPase activity was determined by the procedure of Clarke and Morris [7], protein was measured by the method of Lowry (1951) and SDS-gel electrophoresis was performed according to Laemmli [8].

RESULTS AND DISCUSSION

Growth and specific ATPase activity of B. subtilis decreased when cells were grown in the presence of different concentrations (5,10,15,25 µg/ml) of cerulenin (Figure 1). The decrease in enzyme activity was confirmed using ultracytochemical methods. In cells grown with cerulenin the deposition of the reaction product (lead phosphate) which is usually well-visualized in the cytoplasmic membrane and the mesosomes [5] (Figure 2, A) has been either missing or found in trace amounts (Figure 2 B,C,D). The drug did not affect the electrophoretic patterns of membrane proteins. The ATPase activity of isolated membranes treated in vitro with the antibiotic was the same as that of the controls (not treated). The ultracytochemical studies confirmed these results. Since cerulenin did not cause damage to enzyme activity and to the morphology of the membranes when the latter were

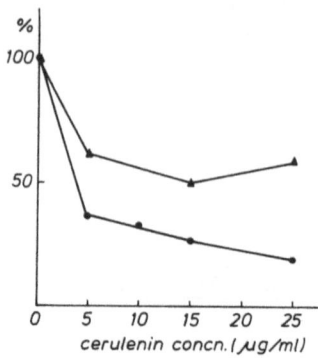

Fig. 1. Effect of cerulenin on B. subtilis 168. ● O.D. at 590 nm (% of a
   control culture without cerulenin); ▲ specific ATPase activity of
   membranes isolated from cells grown in the presence of cerulenin
   (in % of the control).

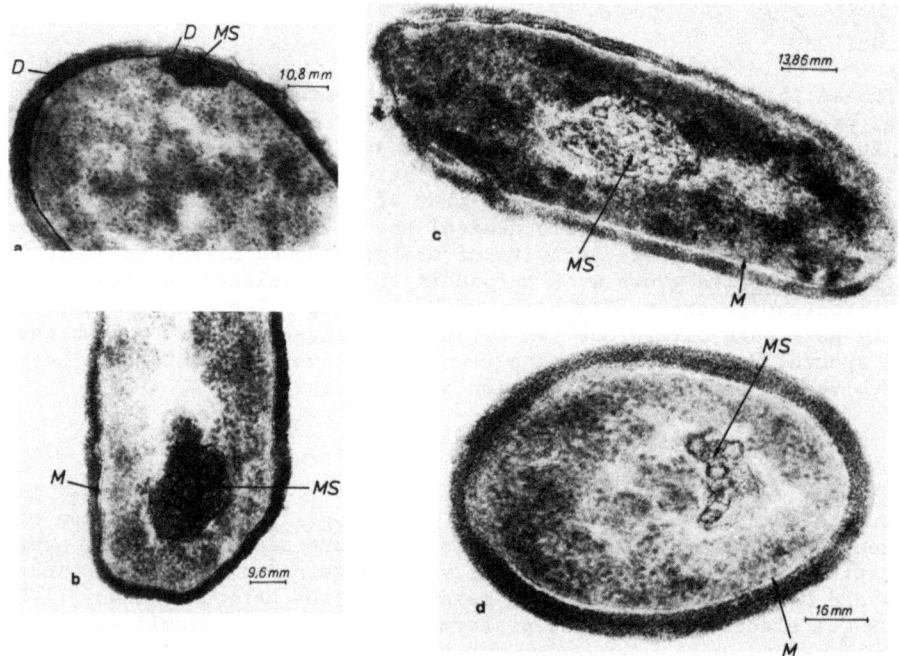

Fig. 2. Ultracytochemical localization of ATPase in B. subtilis 168. A.
   without cerulenin; B. grown with 5µg/ml cerulenin; C. with 15µg/ml
   cerulenin; D. with 25µg/ml cerulenin. Bars, 0.1µm, cytoplasmic
   membrane; MS, mesosomes; D. dense granules.

treated in vitro, it is possible that the agent exercised its action by
changing the lipid composition of the membranes which leads to disorgan-
ization of lipid-depending events in the cell. Thus it cannot be excluded
that cerulenin was able to prevent exocellular enzyme synthesis observed by
several authors by delaying the growth of the cells (it is well-known that
secretion of some enzymes occurs in a definite moment of growth cycle), or
by interfering with the function of the ATPase, suspected to participate in
protein translocation [4].

REFERENCES

1. J. C. Paton, B.K. May, and W.H. Elliott, Cerulenin Inhibits production of extracellular proteins but not membrane proteins in Bacillus amyloliquefaciens, J. Gen. Microbiol., 118:179 (1980).
2. M. F. Petit-Glatron and R. Chambert, Levansucrase of Bacillus subtilis: Conclusive evidence that its production and export are unrelated to fatty-acid synthesis but modulated by membrane-modifying agents, Eur. J. Biochem., 119:603 (1981).
3. P. Mäantsälä, Inhibition of protein secretion by cerulenin in Bacillus subtilis, J. Gen. Microbiol., 128:2967 (1982).
4. L. L. Randall, Function of protonmotive force in translocation of protein across membranes, Meth. Enzymol., 125:129 (1986).
5. N. Cherepova, S. Baykousheva, and K. Ilieva, Ultrastructure of Bacillus subtilis 168 treated with cerulenin, Microbios Lett., 28:127 (1985).
6. N. V. Cherepova, S.P. Baykousheva, and K.Z. Ilieva, Ultracytochemical localization of ATP-hydrolysing activity in vegetative cells, spores and isolated cytoplasmic membranes of Bacillus subtilis 168, J. Gen. Microbiol., 132:669 (1986).
7. D. J. Clarke and J.G. Morris, Partial purification of a dicyclohexyl-carbodi-imide-sensitive membrane adenosine triphosphatase complex from obligately anaerobic bacterium Clostridium pasteuranium, Biochem. J., 154:725 (1976).
8. U. K. Laemmli, Cleavage of structural proteins during the assembly of the head of bacteriophage T4, Nature, London, 227:680 (1970).

PRODUCTION OF EXTRACELLULAR ENZYMES BY

IMMOBILIZED MICROORGANISMS

H. Ruttloff, D. Körner and A. Leuchtenberger

Academy of Sciences of GDR, Central Institute of
Nutrition in Potsdam-Rehbrücke, Bergholz-Rehbrücke
German Democratic Republic

For nearly three decades microbial enzyme preparations have been
increasingly used as technological resources in food production, medicine
as well as in other branches of national economy (Ruttloff et al., 1978;
Ruttloff, 1981, 1986). Numerous enzymes are produced industrially in tons,
especially hydrolases, e.g. amylases, saccharidases, proteinases, pectinolytic enzymes, cellulases, lipases and others. They are still mostly used
in a soluble form.

In the sixties diverse procedures for the immobilization of enzymes
were developed. The advantages of this so-called "2nd enzyme generation"
are obvious (Table 1). The biocatalysts may be used repeatedly, do not
pass into the final product, may be used for continuous processes and have
an enormously increased stability. In the seventies it was assumed that
immobilized enzymes would acquire considerable importance in food production (Table 2). Very soon, however, it became obvious that with regard
to economy only a very few enzymes could be used in immobilized form.
These are glucose-isomerase, lactase and aminoacylase. In the GDR
specially immobilized glucoamylase and invertase are used on an industrial
scale.

Immobilized enzymes found their most important application in biochemical analysis and medicine (diagnostics, therapy). Immobilized enzymes
can also be used for the production of certain compounds via biotrans-
formation (e.g. malic acid, aspartic acid, alanine, etc.); in most cases
they have been replaced by immobilized cells.

The immobilization of cells introduced in the seventies used the same
preparation techniques as those used with enzymes (Table 3), because
according to Chibata (1983) "immobilized cells in the same manner as
immobilized enzymes are physically or chemically confined or localized in a
certain defined region of space with retention of their catalytic activities, and which can be used repeatedly and continuously". These microbial
systems are quasi-immobilized enzymes. However, the latter are localized
within the cell and thus subject to a special protection. There are no
losses of activity during isolation and preparation, especially in the case
of intra-cellular enzymes. Principally we have to differentiate between
two borderline cases. In the first preferably only a single enzyme is
active (e.g. aspartase); the cell is damaged or dead (Table 4). In the
second case several biocatalysts or the complete enzyme system of the cell

Table 1. Advantages and Disadvantages of Enzyme Immobilization (Uhlig, 1984)

Advantages	Disadvantages
Continuous process possible	Losses of enzyme activity with immobilizing process
Improved product quality	
Re-use of enzyme	Leaching effects
Product free of enzyme	Expenses for carrier substances and immobilizing
Inactivation of enzyme (e.g. heat) not necessary	Raised expenditure for equipment and technics
Immobilization of multistep enzyme process possible	Danger of microbial contamination

Table 2. Possibilities for the Application of Immobilized Enzymes in Food Industry (Hartmeier, 1977)

Enzyme	Application
Glucoamylase	Starch conversion
Glucose isomerase	Production of isomerate sugar (HFCS)
Invertase	Production of invert sugar
Lactase	Hydrolysis of lactose (milk, whey)
β-Amylase	Production of maltose
β-Glucanase	Viscosity reduction of wort and beer
Pectinase (polygalacturonase/ pectin esterase)	Clarification of fruit juices
Protease	Chill proving of beer; production of protein partial hydrolysates
Rennin	Cheese production
Glucose oxidase + catalase	Oxygen removal from beverages
Glucose isomerase + glucose oxidase	Production of gluconic acid and frutose from glucose
Catalase	Elimination of H_2O_2 after chemical sterilization
Amino-acylase	Production of L-amino acids from racemic mixtures

are active and the entire metabolism is operative. The vitality of the cell is maintained. i.e. it is intact. The preparation technique in this case is chiefly the embedding of the cells in gels; in the hollow spaces of the gel the cells are able to multiply and attain a rapid increase in the population. This technique is thus likely to be introduced on a large scale.

Table 3. Methods for Cell Immobilization

Carrier binding methods
    Covalent binding
    Ionic binding
    Adsorptive binding

Cross-linking methods
    Application of bifunctional agents
    Application of multifunctional agents

Entrapping methods
    Involvement in a gel matrix
    Encapsulation in a semipermeable polymer membrane

Heating of the biomass

The large-scale production of immobilized enzymes and cells included several stages. As a first step, e.g., glucose isomerase, aminoacylase, β-galactosidase, glucoamylase and invertase were immobilized. The second step was the immobilization of cells with preferably a single enzyme activity (e.g. E. coli = aspartase; Brevibacterium flavum = fumarase; Pseudomonas dacunhae = aspartate decarboxylase). As a third step cells were immobilized but maintained their full vitality and complete enzyme systems. In the near future ethanol (Table 5), edible organic acids, vitamins, amino acids, steroids and other compounds will be produced by means of immobilized cells via total synthesis. (In this connection it may be mentioned that the classical production of vinegar on beech chips in principle represents the use of immobilized microorganisms.) The method of immobilization of cells has spread also to plant and animal cells.

For the purpose of immobilization the microorganisms have to be cultivated, separated, washed, etc., and then submitted to the immobilization step. Losses in activity may occur. Population growth can take place after an immediate, i.e. direct immobilization of the microorganisms and a suitable carrier. An appropriate material is needed to tightly attach the cells or the mycelium to the support (e.g. if the population is growing with hyphae). The surface of the carrier must be rapidly overgrown with no release of the developing cells or hyphae. In this case the mycelium may be considered as "quasi-immobilized" and can be compared to conventionally immobilized microorganisms. However, the enzyme system is not immobilized and it may pass through the cell wall.

Within the last years pectinolytic enzymes have become increasingly important (Figure 1). This enzyme system, composed of endo-polygalacturonase (EC 3.2.1.15.) and pectin esterase (EC 3.1.1.11.) causes a disintegration of the plant tissue by decomposing the middle lamella. Fruit and vegetable cells are separated from each other but remain intact (Figure 2). So-called monocell suspensions are obtained; they have a high nutritional value. Viscosity and stability of the products are maintained if the degradation process is stopped at the proper moment. The ultimate goal is a total liquefaction of plant tissues (Table 6). This task requires studies of the necessary enzyme spectrum and activities, especially with respect to pectinolytic and cellulolytic enzymes.

Pectinolytic enzymes have been conventionally produced by molds, usually in surface culture. We tried to develop a submerged procedure. First experiments with Aspergillus niger demonstrated that the intensity of polygalacturonase production depends in high degree on the morphological

Table 4. Immobilized Cells (avital) with Single Enzyme Activity.

Effective enzyme	Strain	Substrate	Product
Glucose isomerase (Xylose isomerase); EC 5.3.1.5	Bacillus coagulans; Arthrobacter spec.; Streptomyces olivaceus; Streptomyces phaeochromogenes	Glucose	High fructose corn syrup (HFCS)
Invertase; EC 3.2.1.26	Saccharomyces cerevisiae	Sucrose	Invert sugar
β-Galactosidase; EC 3.2.1.23	Saccharomyces fragilis	Lactose	Glucose + galactose
α-Galactosidase; EC 3.2.1.22	Absidia spec.	Raffinose	Sucrose + galactose
Aspartase; EC 4.3.1.1	Escherichia coli	Fumaric acid (NH_4-salt)	L-Aspartic acid
Fumarase; EC 4.2.1.2	Brevibacterium ammoniagenes; Brevibacterium flavum	Fumaric acid	Malic acid
Aspartate-4-decarboxylase; EC 4.1.1.12	Pseudomonas dacunhae	L-Aspartic acid	L-Alanine
Phenylalanine ammonialyase; EC 4.3.1.5	Rhodotorula gracilis	Cinnamic acid (NH_4-salt)	L-Phenylalanine

Table 5. Ethanol Producing Cells (vital)

Effective enzyme system	Strain	Substrate
System of alcoholic fermentation	Saccharomyces cerevisiae; Saccharomyces carls-bergiensis	Glucose, sucrose
Inulase + system of alcolholic fermentation	Kluyveromyces marxianus	Inulin
β-Galactosidase + system of alcoholic fermentation	Kluyveromyces fragilis	Whey (lactose)

Fig. 1. Enzymes acting on the pectin molecule.

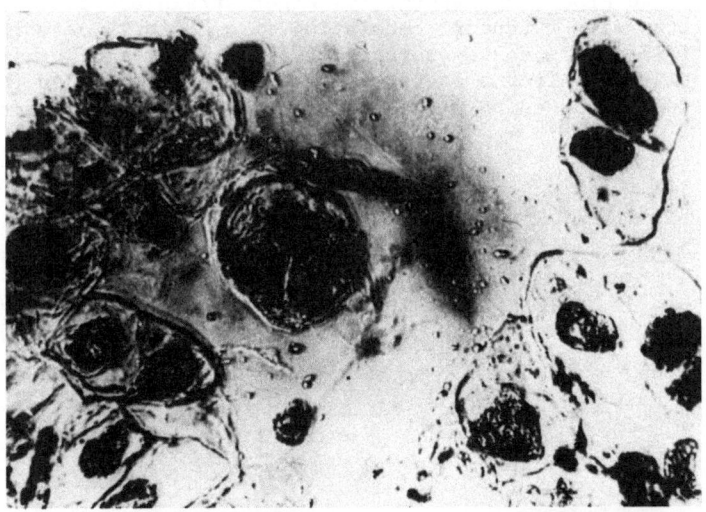

Fig. 2. Carrot tissue after enzymatic maceration.

state of the mycelium. In a shaken culture the enzyme synthesis is much
higher in the presence of compact ball structures ("macro pellets", Figure
3) than with small pellets ("micro pellets", Figure 4) or with hyphae
distributed diffusely in the medium. The cause of this phenomenon (Figure
5) is probably limitation of the transfer of nutrients and gases (O_2, CO_2)

Table 6. Degradation of Plant Tissue by Pectinolytic Enzymes (Bock, 1978)

Enzymes and effects	Biochemical reaction	Desirable process
Pectinolytic enzymes: tissue softening	Partial decomposition of unsoluble protopectins	Natural ripening of fruits and vegetables
Pectinolytic enzymes: tissue loosening	Partial destruction of cell tissue	Juice production; production of carrot juice
Pectinolytic enzymes: tissue maceration	Destruction of the middle lamella; crude fibre fraction as well as cell substances are intact	Production of drinkable mono-cell suspensions or macerates
Pectinolytic + cellulolytic enzymes: total liquefaction	Destruction of the cell wall; liberation and partial destruction of cell substances	Total liquefaction of the plant tissue; production of drinkable hydrolysates

within the compact structure as well as a diminished supply and removal of anabolic and catabolic products. This limitation probably causes the pathologically high biosynthesis of polygalacturonase. Comparison of results with different mycelial forms showed a similar high polygalacturonase biosynthesis by carrier fixed mycelium when the hyphae had a compact structure as in the pellets.

A suitable procedure was developed in our laboratory. Conidia of a polygalacturonase producing Aspergillus niger strain are inocculated into a fermentor fitted with a support consisting of a suitable network (Figure 6). The conidia adhere to the surface of the carrier and germinate quickly and quantitatively. After a short time a surface mycelium is formed which is firmly fixed to the support and can be used for several fermentations.

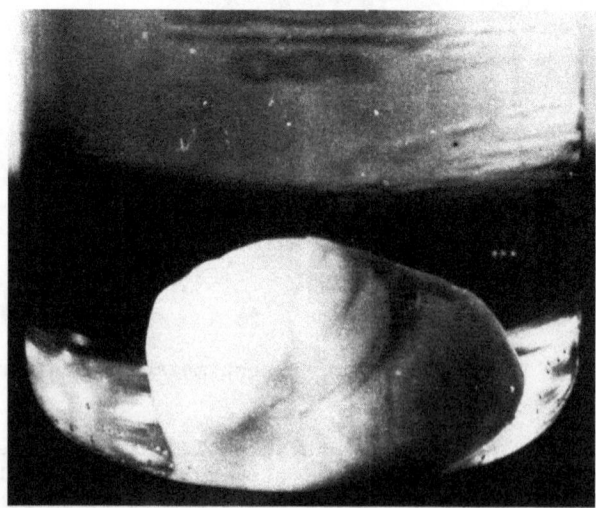

Fig. 3. Big ball structure of mycelium in a shaken culture ('macro pellets') (Hermersdörfer et al., 1980).

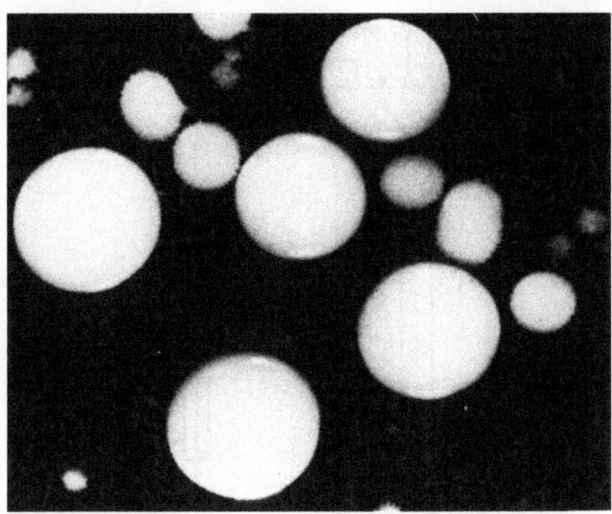

Fig. 4. Small ball structure of mycelium in a shaken culture ('micro pellets') (Hermersdörfer et al., 1980).

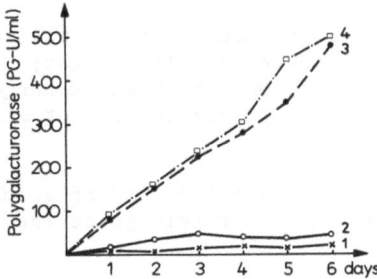

Fig. 5. Polygalacturonase activity of a shaken culture medium with different forms of A. niger mycelium. 1: diffuse mycelium; 2: pellet mycelium ('micro pellets'); 3: big ball mycelium ('macro pellets') from a compact surface mycelial layer; 4: big ball mycelium ('macro pellets') from a compressed diffuse mycelium (Leuchtenberger et al., 1980).

This quasi immobilized mycelium can be compared with a pellet-formed mycelium which, in its very compact form, synthesizes much higher yields of pectinolytic enzymes than a mycelium growing in a diffuse form.

Hence three factors come into effect in the reactor:

a) The "ball effect" occurring in a special shaken culture leads to an increased production of polygalacturonase activity.
b) The reactor works with a quasi-immobilized mycelium.
    Simultaneously it represents a submerged procedure for the production of pectinolytic enzymes. (Submerged pectinase production is thought to be inferior to surface culture especially with A. niger.)
c) The enzymes are not fixed; they may penetrate into the medium.

An additional variant of this technology is the use of enzyme-producing biomass in the fermentor for product processing. An example for this is the clarification of apple juice: turbid apple juice passing

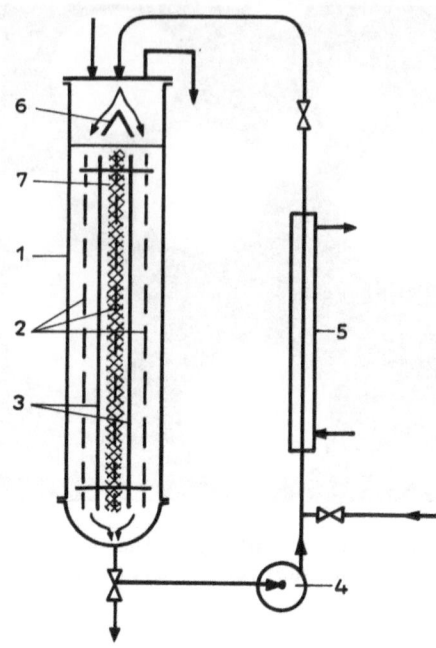

Fig. 6. Laboratory tube reactor with quasi-immobilized mycelium for
polygalacturonase production. 1: reactor; 2: support; 3:
separating plate; 4: pump; 5: heat exchanger; 6: distributor
for medium; 7: mycelium layer (Körner et al., 1984).

through the reactor is rapidly clarified, and if a small quantity of this
clarified apple juice is added to a turbid juice the latter is also rapidly
clarified by sedimentation.

REFERENCES

Bock, W., 1978, Anwendung von pektinolytischen und cellulolytischen Enzymen
    in der Lebensmittelproduktion, Lebensm. Ind., 25: 27
Chibata, I., 1983, Industrial production of optical active compounds using
    immobilized biocatalysts, Basis life Sci., 25: 465
Hartmeier, W., 1977, Immobilisierte Enzyme für die Lebensmittelindustrie,
    Gordian, 77: 202, 232.
Hermersdörfer, H., Wardsack, Ch., Leuchtenberger, A., and Ruttloff, H.,
    1980, unpublished information.
Körner, D., Schiweck, E., and Ruttloff, H., 1984, unpublished information
Leuchtenberger, A., Wardsack, Ch., and Ruttloff, H., 1983, unpublished
    information.
Ruttloff, H., ed., 1981, "Mikrobielle Enzymproduktion", Akademie-Verlag,
    Berlin.
Ruttloff, H., 1986, Impact of biotechnology on food and nutrition, in:
    'Impact of Biotechnology on Food Production and Processing', D.
    Knorr, ed., Marcel Dekker, Inc., New York.
Ruttloff, H., Huber, J., Zickler, F., and Mangold, K.-H., 1978, "Indus-
    trielle Enzyme", Fachbuchverlag, Leipzig.
Uhlig, H., 1984, Einsatz von Enzymen in der Lebensmitteltechnik, Swiss
    Food, 7-8:25.

NEW GENE-TRANSFER SYSTEMS FOR ASPERGILLUS NIGER

Ineke Mattern, Wim van Hartingsveldt, Cora van Zeijl,
Peter Punt, Richard Oliver*, Marianne Dingemanse,
Peter Pouwels and Cees van den Hondel

Medical Biological Laboratory TNO, P.O. Box 45
2280 AA RIJSWIJK, The Netherlands
* School of biological sciences, University of East Anglia
Norwich NR4 7TJ, United Kingdom

INTRODUCTION

The filamentous fungus Aspergillus niger produces a number of extracellular enzymes as well as primary and secondary metabolites, which are of
considerable industrial interest. To facilitate molecular-genetic studies
on the regulation of gene expression and protein secretion in this organism, the availability of an efficient gene-transfer system is of paramount
importance. In this paper we describe the development of two such systems.
The first one is a homologous transformation system, with a vector containing the A. niger pyrG gene as a selectable marker. In the second system, a
vector containing the hygromycin B resistance gene as a dominant selectable
marker is used for transformation. This system is also applicable to
A. nidulans.

RESULTS

Homologous Transformation of A. niger

To develop a homologous transformation system, we have isolated a pyrG
mutant of A. niger and the A. niger pyrG gene (coding for the enzyme
orotidine-5'-phosphate decarboxylase).

Isolation of an A. niger pyrG mutant. A pyrG mutant of A. niger (ATCC
9029) was obtained by UV-irradiation and subsequent selection for resistance to 5-fluoro-orotic acid, according to a procedure developed by Boeke
et al (1984) for Saccharomyces cerevisiae. The putative A. niger pyrG
mutant, AB4.1, could be transformed to Pyr$^+$ with the equivalent pyr^4 gene
of N. crassa, at a frequency of 2 transformants/µg DNA, indicating that
AB4.1 is indeed mutated in the pyrG gene.

Construction of the vector pAB4-1. The A. niger pyrG gene was isolated from a phage lambda genomic library, by screening with a DNA fragment
containing the equivalent pyr^4 gene of N. crassa (Buxton and Radford, 1983)
as a probe. A 3,8 kb XbaI fragment, containing the entire A. niger pyrG

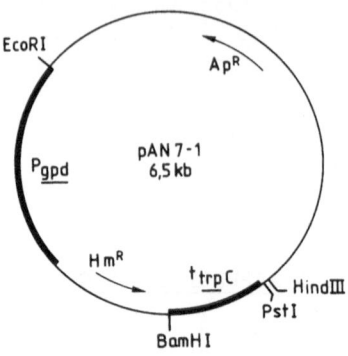

Fig. 1. Plasmid pAB-1. This line: E. coli sequences; thick line: A. niger
        sequences.

gene was subcloned into plasmid pUC19 (Yanisch-Perron et al., 1985),
resulting in the vector pAB4-1 (Figure 1).

   Transformation. A. niger AB4.1 was transformed with pAB4-1, according
to the method described by Yelton et al. (1984) for A. nidulans. Pyr$^+$
transformants were obtained at a frequency of up to 100 transformants/µg
DNA (Table 1). These transformants are mitotically stable. The vector
pAB4-1 is also capable to transform a pyrG mutant of A. nidulans, at a
frequency of 25 transformants/µg DNA.

   Cotransformation. To determine whether pAB4-1 can be used for the
introduction of non-selectable genes into A. niger, AB4.1 was transformed
with a 1:1 mixture of the plasmids pAB4-1 and pAN5-41B (Van Gorcom et al.,
in press). The latter plasmid contains the E. coli lacZ gene, coding for
the enzyme β-galactosidase, fused with the promoter region of the A
nidulans gpd gene, coding for glyceraldehyde-3-phosphate-dehydrogenase.
About 10% of the selected Pyr$^+$ transformants also showed the E. coli
β-galactosidase activity, demonstrating cotransformation of the non-
selected plasmid, and expression of the lacZ gene under control of an
A. nidulans promoter.

   Analysis of integration events. To analyse the integration events
involved in the homologous transformation of A. niger AB4.1 with the vector
pAB4-1, chromosomal DNA of 16 transformants and of the recipient strain was
subjected to Southern blot analysis (Maniatis, 1982). Three types of
integration were found: I. in 8 transformants, replacement of the mutated
pyrG gene by the wild type allele had occurred, II. in 6 transformants 1 or
2 (tandem) copies of the vector had been integrated at the pyrG locus, III.
in 2 transformants the vector had been integrated at a chromosomal site
other than the pyrG locus. Thus, in this homologous transformation a high
frequency (90%) of interaction at the resident pyrG locus was found.

Table 1. Transformation Frequencies of A. niger. AB4.1 and
          A. nidulans G191 (transformants/µg DNA)

Plasmid	Marker	host	
		A. niger	A. nidulans
pAB4-1	pyrG	100	25
pDJB2	pyr4	2	>200
pUC19	–	0	0

52

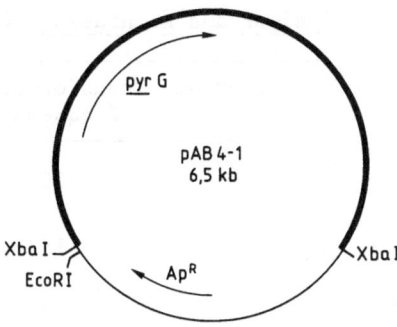

Fig. 2. Plasmid pAN7-1. Thin line: E. coli sequences; thick line: A. nidulans sequences. Sites of EcoRI and PstI are not unique.

Transformation Based on a Dominant Selectable Marker

To facilitate transformation of wild type Aspergillus strains we have developed a gene-transfer system based on the hygromycin B phosphotransferase gene (hph) from E. coli as a dominant selectable marker. Since we found that A. niger and A. nidulans are sensitive to this antibiotic, we have used the cloned E. coli gene (Gritz and Davies, 1983) for the construction of a transformation vector for Aspergillus.

Vector construction. To allow expression in Aspergillus, the hph gene was inserted into the A. nidulans expression vector pAN52-1 (Punt el al., in preparation), resulting in plasmid pAN7-1 (Figure 2). This insertion was carried out in such a way that the hph gene was placed under the control of the promoter region of the A. nidulans gpd gene (van den Hondel et al, 1986), and of the terminator region of the A. nidulans trpC gene (Mullany et al, 1985).

Transformation. Wild type A. niger and A. nidulans strains were transformed with pAN7-1, according to the procedure of Yelton et al. (1984). Transformants, resistant to hygromycin B, were selected on minimal medium agar plates which were incubated at 30-37°C for 16-20 h and subsequently overlaid with agar containing 100 μg/ml (A. niger) or 1000 μg/ml (A. nidulans) of hygromycin B. Transformants appeared after 2-3 days of further incubation at a frequency of 5-15/μg DNA (Table 2).

Cotransformation. To test the possibility of introducing non-selectable genes by cotransformation with pAN7-1, A. niger AB4.1 and an A.nidulans amdS mutant were transformed with a 1:1 mixture of pAN7-1 and pAB4-1 or p3SR2 (Hynes et al., 1983) containing the A. niger pyrG and A. nidulans amdS genes respectively. As shown in table 2, a similar transformation frequency was found for the mutant and the wild type strains. Furthermore, a high efficiency of cotransformation (90-100%) of both unselected plasmids was obtained.   •

CONCLUSIONS

(1) An efficient homologous system to introduce genes in A. niger has been developed. The transformation frequency obtained (100 transformants/μg DNA) is 10-50fold higher than that in the recently described heterologous transformation systems for A. niger, using the A. nidulans amdS (Kelly and Hynes, 1985) and argB (Buxton et al., 1985) genes or the N. crassa pyr[4] gene (table 1). The frequency is high enough to isolate genes of A. niger by complementation of the corresponding mutants, as has been described for

Table 2. Transformation of A. niger and A. nidulans with Various Plasmids

Host	Selected marker	Cotransformed marker	Transformation frequency*	Cotransformation frequency
A. niger (wt)	hph(pAN7-1)	-	10	-
A. niger (pyrG)	hph(pAN7-1)	pyrG(pAB4-1)	15	21/24
A. nidulans (wt)	hph(pAN7-1)	-	5	-
A. nidulans (amdS)	hph(pAN7-1)	amdS p3SR2	15	11/11

* transformants per µg of DNA

A. nidulans (Yelton et al., 1985). The high frequency of integration of pAB4-1 at the homologous pyrG locus suggests that in vitro mutated cloned A. niger genes can be reinserted at their original site, which will facilitate molecular-genetic studies on gene expression and protein secretion.

(2) Wild type and mutant A. niger and A. nidulans strains can be transformed with the vector pAN7-1, which contains the hygromycin B resistance gene as selectable marker, at a frequency of 15 transformants/µg DNA. In addition, efficient cotransformation of other plasmids is possible. These findings make the vector pAN7-1 an important tool for molecular-genetic studies in these Aspergillus species, and possibly also for other filamentous fungi which are sensitive to this antibiotic.

REFERENCES

Boeke, J.D., LaCroute, F. and Fink, G.R. 1984, A positive selection for mutants lacking orotidine-5'-phosphate decarboxylase activity in yeast: 5-fluoro-orotic acid resistance, Mol. Gen. Genet., 197:345.
Buxton, F.P. and Radford, A. 1983, Cloning of the structural gene for orotidine-5'-phosphate decarboxylase of Neurospora crassa by expression in Escherichia coli, Mol.Gen.Genet., 190:403.
Buxton, F.P., Gwynne, D.I. and Davies, R.W. 1985, Transformation of Aspergillus niger using the argB gene of Aspergillus nidulans, Gene, 37:207.
Gritz, L. and Davies, J. 1983, Plasmid encoded hygromycin B resistance: the sequence of hygromycin B phosphotransferase gene and its expression in Escherichia coli and Saccharomyces cerevisiae, Gene, 25:179.
Hynes, M.J., Corrick, C.M. and King, J.A. 1983, Isolation of genomic clones containing the amd S gene of Aspergillus nidulans and their use in the analysis of structural and regulatory mutations, Mol. Cell. Biology, 3:1430.
Kelly, M.K. and Hynes, M.J. 1985, Transformation of Aspergillus niger by the amdS gene of Aspergillus nidulans, EMBO J., 4:475.
Maniatis, T., Fritsch, E.F. and Sambrook, J. 1982, "Molecular cloning- A laboratory manual", Cold Spring Harbor Laboratory, Cold Spring Harbor, NY.
Mullaney, E.J., Hamer, J.E., Yelton, M.M. and Timberlake, W.E. 1985, Primary structure of the Aspergillus nidulans trpC gene, Mol. Gen. Genet., 199:37.

Van den Hondel, C.A.M.J.J., Punt, P.J., Jacobs-Meijsing, B.L.M., van
       Hartingsveldt, W., van Gorcom, R.F.M. and Pouwels, P.H. 1986,
       Analysis of transcription-control signals in _Aspergillus_, _in_:
       "Biology and Molecular Biology of Plant-Pathogen Interactions, NATO:
       1985", J.A. Bailey, ed., Plenum Press, NY in press.
Van Gorcom, R.F.M., Punt, P.J., Pouwels, P.H. and Van den Hondel, C.A.M.J.
       J. 1986, A system for the analysis of expression signals in Asper-
       gillus, _Gene_, in press.
Yanisch-Perron, C., Vieira, J. and Messing, J. 1985, Improved M13 phage
       cloning vectors and host strains; nucleotide sequences of the M13
       mp18 and pUC19 vectors, _Gene_, 33:103.
Yelton, M.M., Hamer, J.E. and Timberlake, W.E. 1984, Transformation of
       Aspergillus nidulans using a _trpC_ plasmid, _Proc. Natl. Acad. Sci._
       USA, 81:1470.
Yelton, M.M., Timberlake, W.E. and van den Hondel, C.A.M.J.J. 1985, A
       cosmid for selecting genes by complementation in _Aspergillus_
       nidulans: selection of developmentally regulated yA locus, _Proc._
       Natl. Acad. Sci. USA, 82:834.

PART II
PROTEINASES AND PEPTIDASES

STRUCTURE OF THERMITASE, A THERMOSTABLE SERINE PROTEINASE
FROM THERMOACTINOMYCES VULGARIS, AND ITS RELATIONSHIP
WITH SUBTILISIN-TYPE PROTEINASES

Miroslav Baudyš, Bedrich Meloun, Vladimir Kostka, Gert
Hausdorf*, Cornelius Frömmel* and Wolfgang Ernst Höhne*

Institute of Organic Chemistry and Biochemistry, Czechoslovak
Academy of Sciences, Prague, Czechoslovakia, and *Institute
of Physiological and Biological Chemistry, Humboldt
University, Berlin, GDR

Thermitase (EC 3.4.21.14) is an extracellular thermostable serine
proteinase isolated from Thermoactinomyces vulgaris culture filtrate [1-3].
The enzyme resembles in its characteristics the subtilisins. This is
strongly indicated especially by the structure of the active site peptide
[4]. The enzyme contains one methionine and one cysteine residue; these
two residues are apparently functionally important [4]. Closely related
seem to be the alkaline proteinases of Bacillus cereus and Bacillus thuringiensis [5] and proteinase K from the mold Tritirachium album [6]. Our
interest in structural investigation of thermitase was stimulated by the
problem of its apparently essential cysteinyl residue, by the lack of
information on evolutionary relations between thermitase and other
subtilisin-type enzymes and on the structural basis of the increased
thermostability of thermitase. Another factor not to be neglected was the
practical importance of the enzyme. Since the localization of the
methionine residue was known from one of the early studies we decided to
start our sequence work with the cyanogen bromide digest of the enzyme.

The cleavage at the single methionine residue gave rise to two specific fragments which were resolved on Sephadex G-50. The sequence of the
C-terminal, smaller fragment CB1 was elucidated by the analysis of peptides
isolated from the limited tryptic digest and from the chymotryptic digest
of the fragment (Figure 1). Judging by this short 53-residue sequence
thermitase is a subtilisin-type proteinase showing a high degree of homology with the subtilisins [7]. This high degree of homology is
understandable since the residue localized next to the methionine residue
at which the molecule was cleaved, is the active site serine residue [4].
A sequence highly homologous with this part of the chain also forms the
core of subtilisin BPN' molecule [8].

The large cyanogen bromide fragment was very poorly soluble; we
decided therefore to approach the sequencing of the remaining part of the
molecule by analyzing the tryptic digest of the whole intact enzyme. The
purification of the tryptic peptides is schematized in Figure 2. Primarily
we were interested in the localization of the SH-peptide and used therefore
thiopropyl-Sepharose for its selective isolation [9]. The enzyme was
denatured before being bound in order to ensure an effective binding and
then cleaved with trypsin. The peptides liberated by tryptic cleavage were

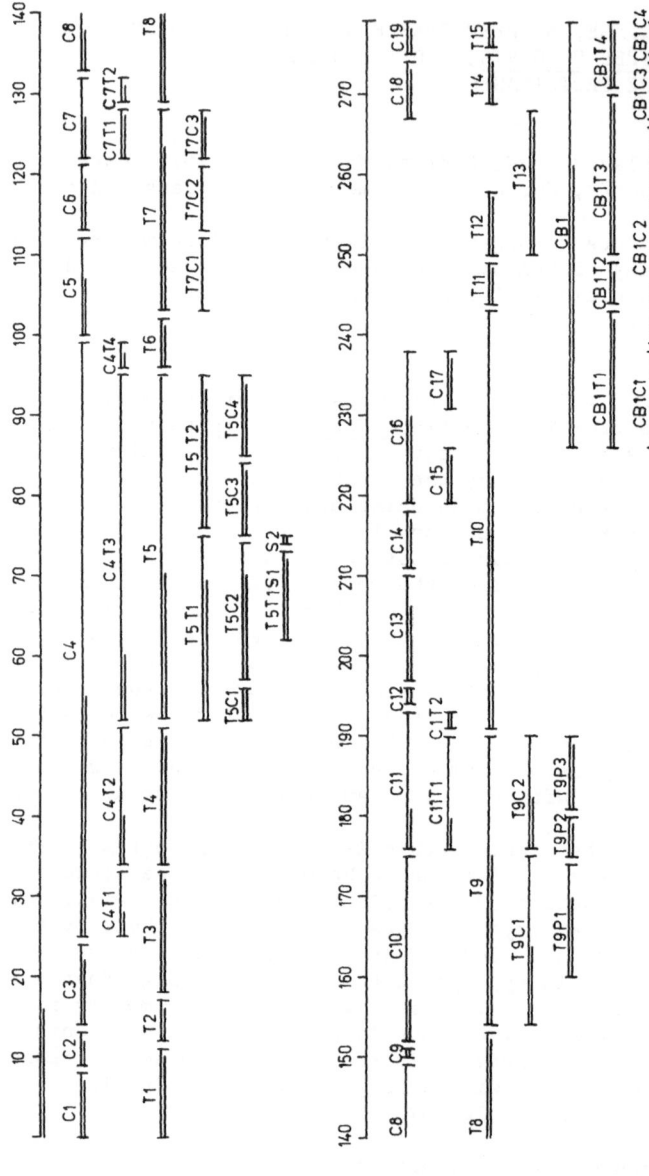

Fig. 1. Alignment of tryptic, chymotryptic and cyanogen bromide fragments and of peptides derived from them with the amino acid sequence of thermitase. The regions sequenced are marked by a full line under the horizontal bars representing the individual peptides. Individual symbols: T — tryptic peptide, C — chymotryptic peptide, S — subtilisin peptide, CB — cyanogen bromide fragment.

eluted by 0.1 M ammonia and the SH-peptide was then selectively displaced from thiopropyl-Sepharose by mercaptoethanol. The unbound peptides eluted by ammonia were lyophilized and the lyophilisate extracted with 0.1% TFA. The soluble peptide fraction was separated by reversed phase HPLC, the insoluble part was solubilized in 50% TFA and the peptides were resolved by gel chromatography on Sephadex G-50 in 30% acetic acid.

The SH-peptide was degraded in a spinning cup sequenator (Figure 1). Because of the repetitive sequence Asn-Gly (Figure 3), the degradation did not proceed too far. We therefore prepared secondary tryptic peptides after amino-ethylation of the cysteinyl residue of the SH-peptide. The parts of the sequence which could not be determined by the analysis of the two tryptic digests were obtained after subtilisin digestion of the SH-peptide. The alignment of the sequence obtained with the sequence of the corresponding parts of the subtilisin molecules (Figure 3) showed that the SH-peptide contains the active His 64 of the subtilisins and that the cysteine residue is localized in its neighborhood. To verify this assumption we inactivated the enzyme with Z-Ala-Ala-Phe- $^{14}CH_2Cl$, which specifically reacts with the active site histidine residue of the subtilisin-type enzymes. All the radioactivity incorporated into the enzyme at a molar enzyme-to-label ratio of 1:1 was recovered in the SH-peptide. Moreover, the labelled peptide contained two more alanine residues than the parent peptide, corresponding to a 1:1 incorporation of the inactivator. As expected, one histidine residue was missing in the amino acid composition of the labelled fragment since the chloromethyl ketone group specifically reacts with the imidazole ring of the active site histidine [10].

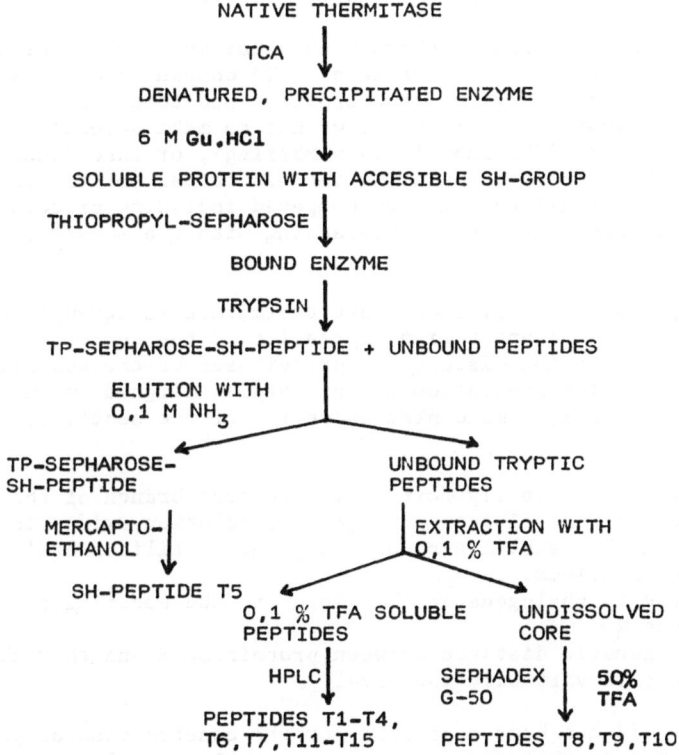

Fig. 2. Flow diagram of purification of tryptic peptides of thermitase. The enzyme was denatured to obtain an efficient attachment. TCA - trichloroacetic acid, Gu.HCl - guanidinium hydrocholoride, TFA - trifluoroacetic acid. See text for details.

Having determined the structure of the SH-peptide we started investigating the fraction of peptides soluble in 0.1% TFA. The peptides were separated by reversed phase HPLC (Figure 4). Save for one exception all the peptides were obtained in pure state and their sequences were established in a solid-phase protein sequencer in the case of peptides with C-terminal lysine, or in the spinning cup sequenator in the case of peptides C-terminated with arginine (Figure 1,3).

The peptides of the insoluble core were solubilized in 50% TFA and resolved on a Sephadex G-50 column. Three peaks were obtained and their rechromatography afforded the remaining tryptic peptides. The sequence of peptide T8 as well as the N-terminal sequences of peptides T9 and T10 were determined in the spinning cup sequenator (Figures 1 and 3). The C-terminal portion of peptide T9 was elucidated by the analysis of peptides derived from its peptic and chymotryptic digest (Figure 1). The analysis of the C-terminal part of the sequence of peptide T10 was unnecessary since it represents the N-terminus of cyanogen bromide fragment CB1 whose complete sequence was known from preceding experiments (see above).

Peptides T1 through T15 account for the entire chain of thermitase and there is a large overlap of peptides T10 and CB1 (Figure 1). The overlaps of the tryptic peptides afforded the investigation of the chymotryptic digest of the whole protein. The same strategy as in the case of the tryptic digest was applied. We selectively isolated the SH-peptide C4 (Figure 1), the peptides not bound to thiopropyl-Sepharose were separated by two-dimensional reversed phase HPLC. The peptides isolated yielded all the overlaps necessary. All these experiments permitted us to derive the complete primary structure of the enzyme (Figure 3).

The polypeptide chain of thermitase consists of 279 residues and hence corresponds well to various subtilisins. To obtain an optimal alignment with the latter the N-terminus of thermitase had to be shifted to the left by 7 residues (Figure 3). Likewise, we had to make several deletions (No. 18, 58, 162-165, 238-239, subtilisin numbering), or insertions (No. 49-50, 83, 260, thermitase numbering) along the middle part of the chain. Identical amino acids of all the enzymes compared including proteinase K are enboxed. S_1-S_4 denote residues interacting with the substrate at the individual subsites [11,12].

The comparison clearly shows that thermitase is strongly homologous with subtilisins. A quantitative expression of the evolutionary relations is given by the mutation distance of proteinases of the subtilisin type [13] (Table 1). The population of the enzymes aligned in the table also involved proteinase K, a mold proteinase [6] of the subtilisin type. These data show that

(a) the subtilisins clearly form an independent branch of the phylogenetic tree and that the closest phylogenetic relation exists in the pairs subtilisin DY - subtilisin Carlsberg and subtilisin BPN' - subtilisin Amylosacchariticus.
(b) thermitase is phylogenetically closer to the subtilisins than to proteinase K
(c) the phylogenetic distance between proteinase K and the subtilisins or thermitase is virtually the same.

Since we did not have a program for the construction of phylogenetic trees at the moment the sequence work was completed [13] we prepared a simplified table of mutation distances (Table 2). We statistically averaged the values for subtilisins, thus forming one subtilisin, and constructed a simplified phylogenetic tree (Figure 5). As expected, subtilisin and thermitase as representatives of the procaryotes form one branch. This

Table 1. Mutation Distances of Serine Proteinases of the Subtilisin Type.

	Thermitase	Subtil. Carblsb.	Subtil. DY	Subtil. Amylos.	Subtil. BPN'	Proteinase K
Thermitase						
Subtilisin Carlsberg	224					
Subtilisin DY	225	44				
Subtilisin Amylosacch.	242	104	103			
Subtilisin BPN'	245	110	111	43		
Proteinase K	301	288	291	275	276	

The mutation distance between two members of the population is defined as
the mininal number of nucleotides which must be altered in order for the
gene encoding one member to encode the other number [13]. Uncommon
deletions are penalized by 3 nucleotide replacements irrespective of the
number of amino acid residues involved.

finding is thus in accordance with the generally accepted evolutionary
relations. The lower part of the figure shows an unsimplified tree, in
this case without the mutation distances; it qualitatively illustrates the
phylogenetic relations among all the members of the group.

In addition to thermitase several other subtilisin-like proteinases
are known at present which also contain a cysteine residue apparently
essential for catalytic activity. This is, for instance, proteinase K in
which the topological equivalence of its cysteine residue with that of
thermitase has been demonstrated [6], or the proteinases from Bacillus
cereus and B. thuringiensis. Fragmentary sequence data on the Bacillus
cereus proteinase are also available (Figure 6) [5]. Figure 6 shows a
comparison of the partial primary structures from the N-terminus of the
molecule and from the neighborhood of the catalytically active serine
residue of these enzymes. If these relations are expressed in quantitative
terms (Table 3), this time in percent of homology among subtilisin BPN',
thermitase and the Bacillus cereus proteinase, a close phylogenetic
relationship between thermitase and the Bacillus cereus proteinase can be
seen. At the same time, however, the genus Actinomyces producing
thermitase and the genus Bacillus producing the remaining two proteinases
of this triad belong to biologically different taxonomic groups. Why
thermitase from Actinomyces resembles the Bacillus cereus proteinase more

Table 2. Simplified Table of Mutation Distances of Subtilisin-type
Proteinases.

	Subtilisin	Thermitase	Proteinase K
Subtilisin			
Thermitase	233		
Proteinase K	283	301	

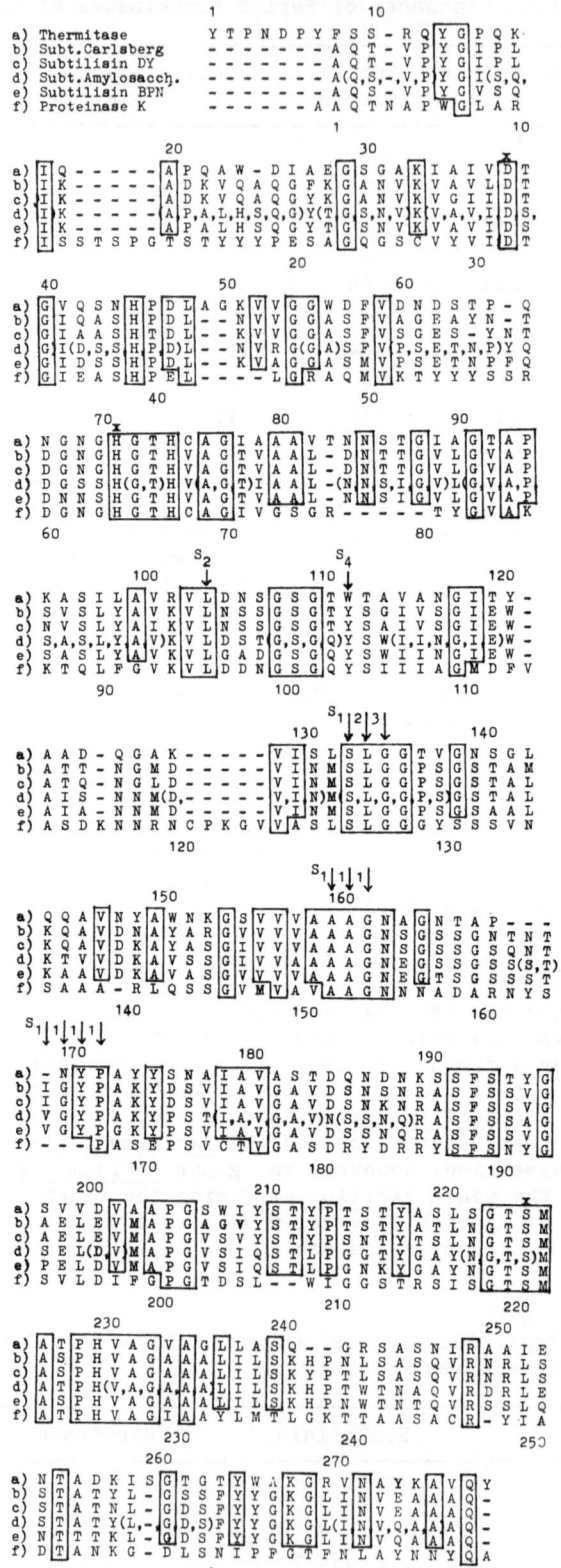

Fig. 3. Comparison of sequences of thermitase, subtilisins and proteinase
K. Top numbering – thermitase, bottom numbering – subtilisin BPN'.
S_1–S_4 denote amino acid residues forming the individual subsites
for the interaction with the substrate [11]. Asterisks mark the
essential amino acid residues of the catalytic triad [8].
Identical residues of thermitase and subtilisins are enboxed;
identical residues of proteinase K are also included. The amino
acids are represented by the one-letter code recommended by IUPAC.
Subt. Amylosacch. – subtilisin Amylosacchariticus.

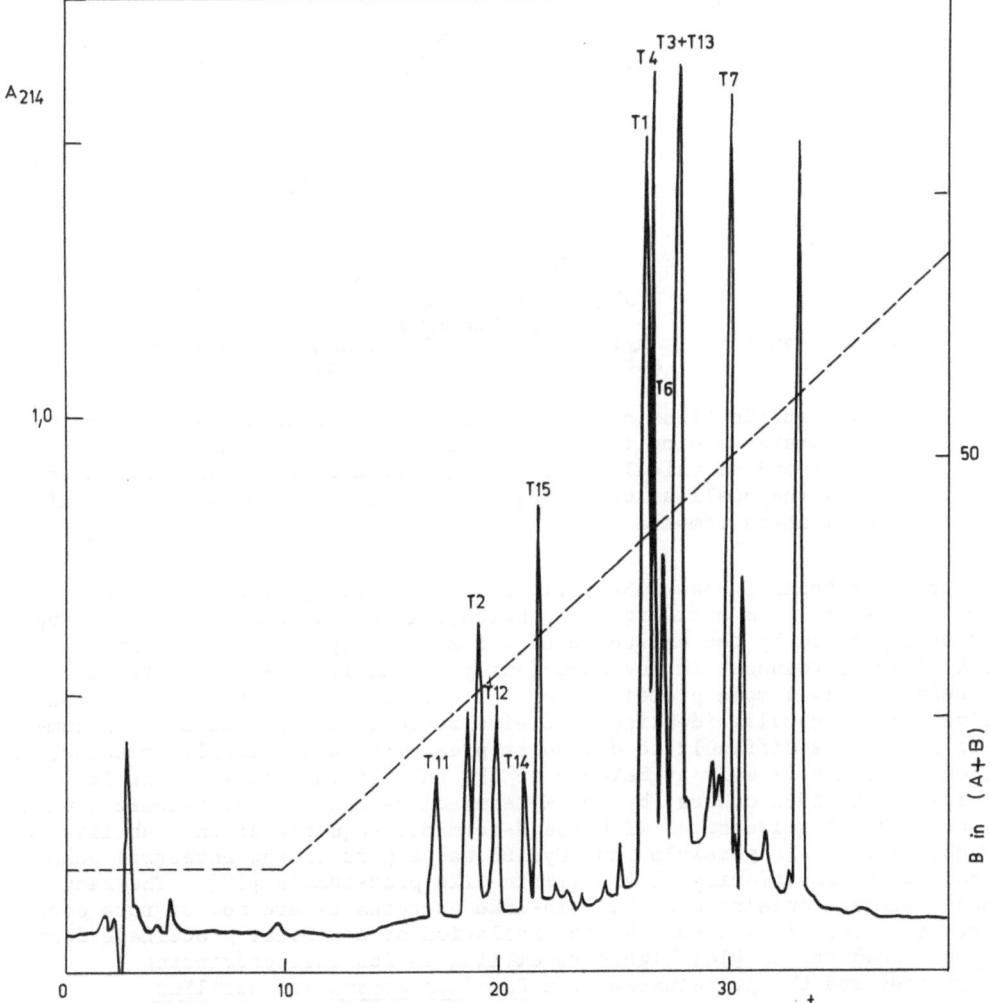

Fig. 4.   Resolution of a mixture of tryptic peptides from thermitase
soluble in 0.1% TFA by HPLC on a reversed phase column. The
lyophilisate (0.5 mg) was extracted with 0.5 ml of 0.1% TFA. The
supernatant was injected into an Ultrasphere C8 column (4.6 mm x
25 cm). Buffer A – 0.1% TFA, buffer B – 0.1% TFA, 80%
acetonitrile. Flow rate 1.25 ml min^{-1}. The gradient profile is
marked by a dashed line. For symbols see Figure 1. t – time (min).

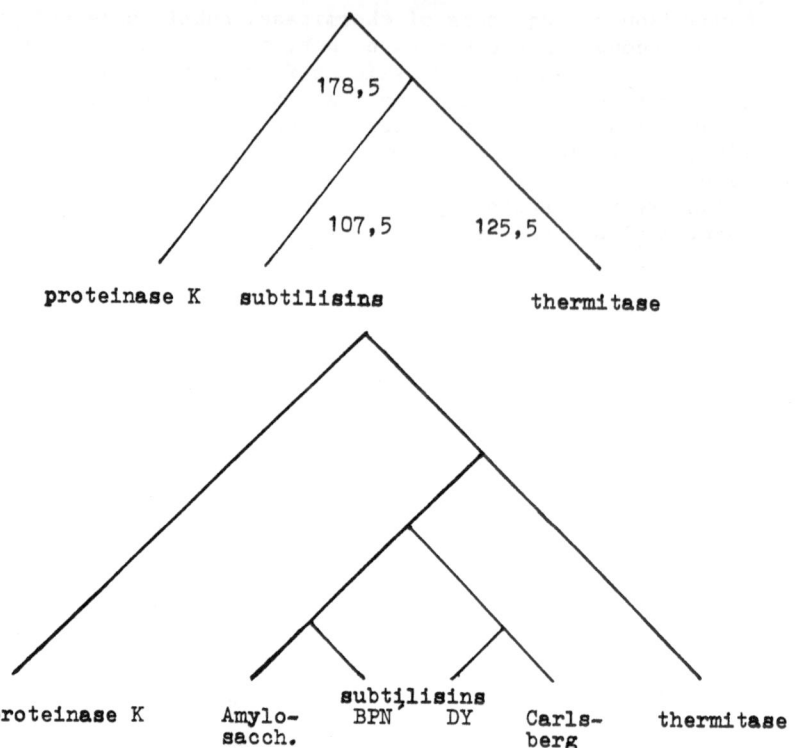

Fig. 5. Phylogenetic trees of subtilisin-type proteinases. The upper part
illustrates a simplified phylogenetic tree. The values given
correspond to calculated mutation distances [13]. The lower part
shows the qualitative phylogenetic interrelationships among all
the members compared.

than this proteinase does subtilisin BPN', even though the latter two
proteinases are from microorganisms belonging to the same taxonomic group,
can be explained by the existence of a family of cysteine-containing
subtilisin proteinases in the superfamily of subtilisin-like proteinases.
Proteinase K is a mold proteinase of the subtilisin type yet it also contains a topologically identical cysteine residue (Cys 68, subtilisin numbering). It is difficult to decide at present to which family proteinase K
belongs but most likely it belongs to the subtilisin family. This is
indicated, besides others, by the N-terminal sequence of proteinase K which
is topologically identical with the N-terminal sequence of the subtilisins
(Figure 3). It is possible that Cys 68 was a part of the ancestral gene
encoding the superfamily of subtilisin-like proteinases [13]. The fact
that cysteine-containing subtilisin-like proteinases are not of rare occur-
rence in nature is evidence by the isolation of a similar proteinase from
Streptomyces rectus [14] highly resembling in its characteristics
thermitase and the proteinases from Bacillus cereus and Bacillus
thuringiensis.

The highest degree of homology of thermitase with subtilisins can be
observed in the neighborhood of the active site of the enzyme, i.e. around
Ser 221, His 64 and Asp 32 (Figure 3). The three-dimensional arrangement
of the catalytic triad of thermitase is obviously very similar to the
spatial configuration of the active residues of subtilisins [8]. Highly
homologous with the subtilisins are also those parts of the chain which
form the substrate-binding subsites S_1-S_4 in subtilisin BPN', as shown by
X-ray analysis of complexes of subtilisin BPN' with various substrate

Table 3. Homologies of Determined Parts of Primary Structures of Subtilisin BPN', Thermitase and Bacillus cereus Proteinase.

	Subtilisin BPN'	Thermitase	Bacillus cereus proteinase
Subtilisin BPN'			
Thermitase	39.5		
Bacillus cereus proteinase	37.2	65.1	

The similarity of the sequences is expressed in percent of homology. This percent is given by the number of identical residues for the given pair divided by the number of residues of the longest chain and multiplied by 100.

analogs [11,12]. The only difference observed within these regions is the deletion of 4 amino acid residues in the thermitase sequence at the edge of one part of the hydrophobic binding pocket forming subsite S_1, and the replacement of Tyr 104 of the subtilisins, which forms subsite S_4, by tryptophan in thermitase. This replacement is very interesting since Tyr 104 has been shown to possess segmental mobility in subtilisin complexes with inhibitors [11]. Hence, the three-dimensional structure of the binding site of thermitase is obviously very similar to that of the subtilisins. Kinetic mapping of the subsites and the investigation of the secondary specificity of thermitase with respect to the subtilisins strongly supports this assumption [15]. The identical residues in other parts of the chain are distributed relatively evenly. The number of structurally conservative replacements for the pair thermitase/subtilisin Carlsberg equals 60, that is more than 20% of residues, a fact indicating a similar three-dimensional structure of these two enzymes. The determination of the surface residues of subtilisin BPN' according to its three-dimensional model using the calculation of exposure indexes [16] showed that 75% of nonconservative amino acid replacements for the pair subtilisin BPN' - thermitase are on the surface of the molecule, mainly in the region of the loops. This holds true in full also for the deletions and insertions found.

```
                              1                    10
Subtilisin BPN´            A Q S V P Y G V S Q I K A
Thermitase        Y T P N D P Y F S S R Q Y G P Q K I Q A
B.c.proteinase    W T P N D P Y Y K N - Q Y G L Q X L X A

14      223           230                 240   243
P A .....A S P H V A G A A A L I L S K H P N W T N ......
P Q .....A T P H V A G V A G L L A S Q G R S A S N ......
P N .....A T P X V A Q V A A L L A N Q G Y S N T Q ......
```

Fig. 6. Comparison of available sequence data on subtilisin BPN', thermitase and Bacillus cereus (B.c.) proteinase. Identical amino acid residues of thermitase and B.c. proteinase are enboxed, identical residues of subtilisin BPN' are also included.

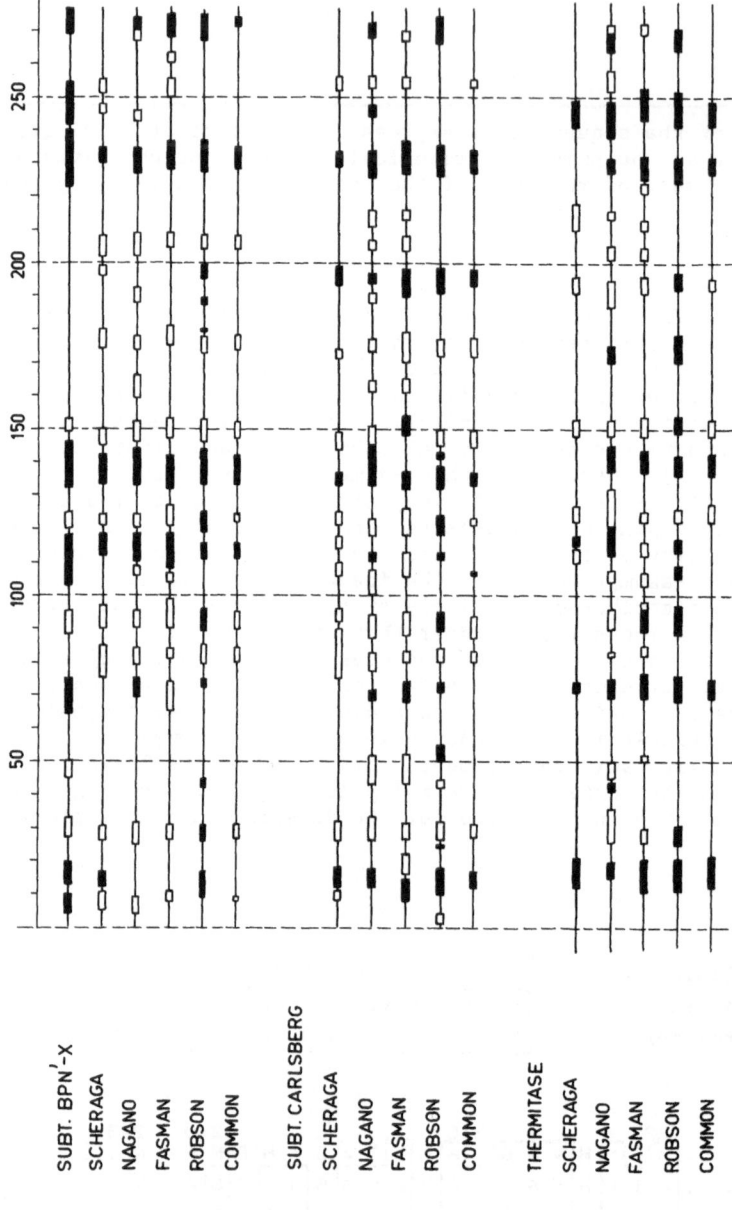

Fig. 7. Prediction of secondary α– ■ and β– ☐ structure of subtilisin BPN', subtilisin Carlsberg and thermitase by 4 independent methods by Scheraga, Nagano and Robson cited in [17]. The final prediction (common) is represented by the agreement of at least three of the four predictions.

For the above mentioned enzymes we have made also a prediction of the secondary structure which underlies the three-dimensional structure (Figure 7). To determine the degree of reliability of the 4 independent methods used [17] we made the prediction of the subtilisin BPN' structure which has also been determined experimentally [8]. The figure shows in all cases only α and β structures. As final prediction is taken the one which represents an agreement of an least three of the four independent predictions. The secondary structure of subtilisin Carlsberg was predicted as an intermediary structure since both subtilisins had been shown to possess identical three-dimensional structures [18]. Better results were obtained with subtilisin BPN', obviously because it is a member of the statistical population used for the calculation of prediction parameters of the methods employed. The differences, however, are not marked. The secondary structure is "underpredicted" in both cases. A slight "overprediction" can be observed in some parts of the chain residues (160-220). From this viewpoint the prediction of the thermitase is equally successful and globally agrees with the secondary structure of subtilisin BPN'. The probability of positive prediction varies around 50%, the overprediction for thermitase is minimal. The percent of successfulness obtained generally agrees with the reported data [19]. A better result could be obtained if the effect of long-range interactions and of hydrophobic effects during folding of the chain could be eliminated. It has been shown that the prediction parameters used so far for the β-structure are applicable better to the buried residues [19], that is to those parts of the chain which constitute the "core" of the molecule. This is in agreement with the positive prediction of three short β-structures (around residues 30, 120 and 150) in all three representatives. The exposition indexes of these regions of subtilisin BPN' are close to zero [16]. The agreement obtained between the predicted secondary structure and the structure determined experimentally for subtilisin BPN' and Carlsberg [8,18], can be regarded as another indirect proof of the homology of thermitase with the subtilisins at the level of three-dimensional structures. We may thus conclude that these structures of thermitase and subtilisin BPN' are very similar. This conclusion receives further support from the data on the three-dimensional structure of proteinase K and of its complex with a low molecular weight inhibitor [12]. The superposition of the equivalent α-carbon atoms of subtilisin BPN' and of proteinase K gives a root-mean-square deviation of only 1 Å. The structure of the active site region of proteinase K gives us also an exact picture of the position of active site His 64 and of the cysteine residue topologically identical with that of thermitase. The cysteine residue is hidden under the imidazole ring of histidine. This is in agreement with the lack of reactivity of the SH-group of cysteine to conventional SH-reagents [4]. Moreover, the exposition index of the topologically identical valine residue in subtilisin BPN', whose side chain is also oriented from the inside toward the imidazole ring of histidine 64, is close to zero [16]. We may adopt the mechanistic view that the modification of cysteine by mercury compounds interferes with the arrangement of the active site which is necessary for effective catalysis by the enzyme.

It has been hypothesized that thermostability is markedly affected by the solvation of hydrophobic surface residues [20]. In spite of the fact that the surface of thermitase is a little more hydrophobic than the surface of the subtilisin BPN' molecule, judging by exposition indexes, thermitase has a higher thermostablity than expected [3].

Another line of approach to the elucidation of increased thermostability of enzymes is the analysis of significant priorities in amino acid replacements, resulting from a comparison of pairs of homologous proteins from thermophiles and mesophiles [21] (Figure 8). The left-hand part of the figure shows a general scheme derived from a statistical treatment of several pairs of thermostable and thermolabile enzymes,

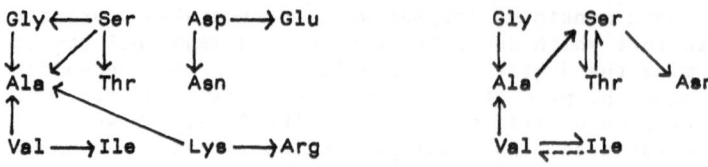

Fig. 8. Left-hand part - general scheme of amino acid replacements in direction of increased thermostability derived for pairs of homologous thermostable and thermolabile enzymes taken from Argos et al. [21] Right-hand part - scheme derived from comparison of amino acid replacements between thermitase and subtilisins BPN' and Carlsberg.

the right-hand part shows a scheme describing the pair thermitase - subtilisin BPN'.

The replacement of Val and Gly by Ala and of Val by Ile corresponds to the general scheme. No replacements of ionized amino acids, however, can be observed. The drawback of all these methods is that they are statistical in nature; they include therefore also the genetic drift which is of more or less neutral character. A change in thermostability can be caused, however, by the replacement of a single amino acid, resulting, for example, in the formation of a new ion pair [22] or in an aromatic-aromatic interaction [23]. The latter type of stabilization can easily play a role in thermitase which, unlike subtilisin BPN', contains four additional tyrosine residued and three additional tryptophan residues. The final answer, however, will have to await the experimental determination of the three-dimensional structure of the enzyme.

REFERENCES

1. C. Frömmel, G. Hausdorf, W.E. Höhne, U. Behnke, and H. Ruttloff, Charakterisierung einer protease aus Thermoactinomyces vulgaris (Thermitase), Acta Biol. Med. Germ., 37:1193 (1978).

2. R. Kleine, Properties of thermitase, a thermostable serine protease from Thermoactinomyces vulgaris, Acta Biol. Med. Germ., 41:89 (1982).

3. C. Frömmel, and W.E. Höhne, Influence of calcium binding on the thermal stability of thermitase, a serine protease from Thermoactinomyces vulgaris, Biochim. Biophys. Acta, 670:25 (1981).

4. G. Hausdorf, K. Krüger, and W.E. Höhne, Thermitase, a serine protease from Thermoactinomyces vulgaris, Int. J. Peptide Protein Res., 15:420 (1980).

5. O. P. Zagnitko, G.G. Chestukhina, L.P. Revina, and V.M. Stepanov, Thiol-dependent serine proteinases, Bioorgan. Khim., 10:383 (1984).

6. K. -D. Jany, G. Lederer, and B. Mayer, Amino acid sequence of proteinase K from the mold Tritirachium album Limber, FEBS Lett., 199:139 (1986).

7. M. Baudyš, V. Kostka, K. Grüner, G. Hausdorf, and W.E. Höhne, Amino acid sequence of the small cyanogen bromide peptide of thermitase, a thermostable serine proteinase from Thermoactinomyces vulgaris, Int. J. Peptide Protein Res., 19:32 (1982).

8. J. Kraut, Subtilisin: X-ray structure, in: "The Enzymes", P.D. Boyer, ed., Acad. Press, New York and London (1971).

9. M. Baudyš, V. Kostka, G. Hausdorf, S. Fittkau, and W.E. Höhne, Amino acid sequence of the tryptic SH-peptide of thermitase, Int. J. Peptide Protein Res., 22:26 (1983).

10. F. S. Markland, and E.L. Smith, Subtilisins: Primary structure, chemical and physical properties, in: "The Enzymes", P.D. Boyer, ed., Acad. Press, New York and London (1971).

11. J. D. Robertus, R.A. Alden, J.J. Birktoft, J. Kraut, J.C. Powers, and P.E. Wilcox, An X-ray crystallographic study of the binding of peptide chloromethyl ketone inhibitors to subtilisin BPN', Biochemistry, 11:2439 (1972).

12. C. Betzel, G.P. Pal, M. Struck, K.-D. Jany and W. Seanger, Active-site geometry of proteinase K, FEBS Lett., 197:105 (1986).

13. W. M. Fitch and E. Margoliash, Construction of phylogenetic trees, Science, 155:279 (1967).

14. K. Mizusawa and F. Yoshida, Thermophilic Streptomyces alkaline proteinase, J. Biol. Chem., 247:6978 (1972).

15. D. Brömme, K. Peters, S. Fink and S. Fittkau, Enzyme-substrate interactions in the hydrolysis of peptide substrates by thermitase, subtilisin BPN' and proteinase K, Arch. Biochem. Biophys., 244:439 (1986).

16. B. Meloun, M. Baudyš, V. Kostka, G. Hausdorf, C. Frömmel, and W.E. Höhne, Complete primary structure of thermitase from Thermoactinomyces vulgaris and its structural features related to the subtilisin-type proteinases, FEBS Lett., 183:195 (1985).

17. M. Dzionara, S.M.L. Robinson and B. Wittman-Liebold, Secondary structure of proteins from the 30S subunit of the Escherichia coli ribosome, Hoppe-Seyler's Z. Physiol. Chem., 358:1003 (1977).

18. W. Bode, E. Papamokos, D. Musil, U. Seemüller, and H. Fritz, Refined 1,2 Å crystal structure of the complex formed between subtilisin Carlsberg and the inhibitor eglin C. Molecular structure of eglin and its interaction with subtilisin, EMBO J., 5:813 (1986).

19. R. C. Garratt, W.R. Taylor, and J.M. Thornton, The influence of tertiary structure on secondary structure prediction, FEBS Lett., 188:59 (1985).

20. E. Stellwagen, Strategies for increasing the stability of enzymes, Ann. N.Y. Acad. Sci., 434:1 (1985).

21. P. Argos, M.G. Rossmann, U.M. Grau, H. Zuber, G. Frankand, and J.D. Tratchin, Thermal stability and protein structure, Biochemistry, 18:5698 (1979).

22. M. F. Perutz, and H. Raidi, Stereochemical basis of heat stability in bacterial ferredoxins and in hemoglobin A2, Nature, 255:256 (1975.

23. S. K. Burley, and G.A. Petsko, Aromatic-aromatic interactions in proteins, Science, 229:23 (1985).

REGULATION OF SYNTHESIS OF AN EXTRACELLULAR PROTEINASE

IN GROWING AND SPORULATING BACILLUS MEGATERIUM

J. Chaloupka, H. Kucerová, M. Strnadová, L. Váchová,
M. Dvoráková, J. Pazlarová, J. Moravcová, J. Votruba
and K. Vanatalu

Depts of Enzyme and Process Engineering, Institute of
Microbiology, Czechoslovak Acad. Sci., Prague 4

Extracellular proteolytic enzymes of microbial origin are often employed in various branches of industry. The neutral and alkaline proteinases of Bacillus subtilis belong to the most extensively used enzymes. However, these proteinases are also of great significance from the theoretical point of view. They are synthesized during the post-exponential and stationary phases of growth and the control of their formation is somehow related to sporogenesis [1,2].

Bacillus megaterium, unlike B. subtilis, produces only one type of the extracellular proteinase - the neutral metalloenzyme - and the enzyme is synthesized and excreted during exponential growth as well as during sporogenesis [3]. The characteristics of the enzymes formed by the exponentially growing or sporulating populations indicate that the proteinases are probably identical [4]. This offers the possibility to compare the regulation of the proteinase during different developmental phases. Moreover, the stability of the metalloproteinase of B. megaterium is strongly dependent on the presence of Ca^{2+}. The enzyme synthesized in its absence is inactivated almost instantaneously. It is therefore possible to follow the rate of the proteinase formation by means of "Ca^{2+} pulses", i.e., by adding Ca^{2+} to a population growing or incubated in its absence and by determining the enzyme activity a few minutes afterwards using a labelled protein as substrate [5].

The extracellular metalloproteinase is almost quantitatively excreted from the cells to the cultivation medium, only a negligible portion remains on the cell walls or in the periplasm and is released during protoplasting of cells with lysozyme. The true intracellular proteinase(s) belong(s) to the serine-type enzymes [5,6].

This contribution summarizes our results of the study of the regulation of the extracellular proteinase. In its first part, we present data concerning the control of the proteinase synthesis during growth of an asporogenic mutant of B. megaterium KM. In the second part, we compare the physiological and biochemical aspects of its regulation during growth and sporulation of a wild strain sporulating with a high efficiency.

The experiments were carried out with cultures growing in a shaken water bath or in laboratory fermenters (2 L). The basic kinetic parameters of proteinase formation were very similar in cultures growing exponentially both in shaken Erlenmeyer flasks and in fermenters.

The population grew in a synthetic mineral salts medium with glucose as carbon source at 35°C with specific growth rate μ = 0.48 (mean generation time 1.4 h) and yield coefficients of the proteinase Y_{DW} = 0.40 and Y_G = 0.13, where Y_{DW} is the enzyme activity in tyrosine units (TU) formed per 1 mg of dry weight increase, and Y_G the activity (TU) produced per 1 mg of glucose consumed. The cultivation in laboratory fermenters proceeded at constant pH (6.5) and air supply (V/V). The exponential growth was paralleled by an exponential production of the proteinase (Figure 1). The relationship between the proteinase synthesis, dry weight increase, glucose and NH_4^+ consumption shows that the increase of enzyme activity is proportional to the increase in culture dry weight and is not affected by the decreasing concentrations of either glucose or ammonia (Figure 2). This indicates that neither catabolite repression nor repression with NH_4^+, which are involved in the control of several microbial proteinases, play any significant role in the regulation of the metalloproteinase of B. megaterium.

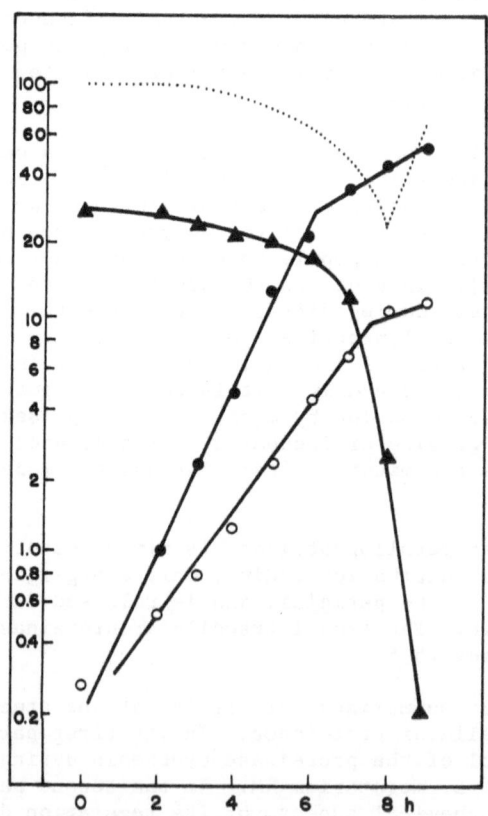

Fig. 1.   Growth and proteinase formation in a laboratory fermenter at 35°C. Abscissa–time in hours; ordinate: empty circles – dry matter (mg/mL), full circles – proteolytic activity (TU/10 mL), full triangles – glucose (mg/mL), dotted line – oxygen saturation (%).

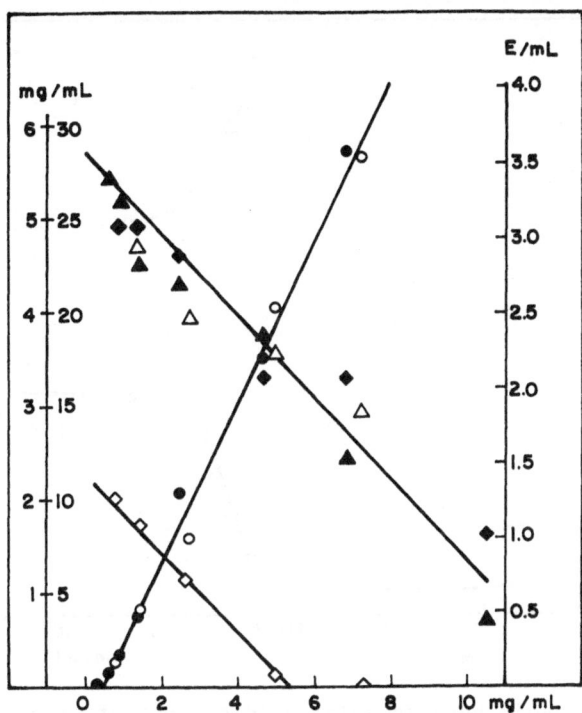

Fig. 2. Relationship between dry matter and proteinase increase and
substrate utilization. Abscissa: dry matter (mg/mL); ordinate:
circles, E/mL - proteolytic activity (TU), triangles, mg/mL -
glucose, diamonds, mg/mL - ammonium chloride; empty symbols -
ammonium chloride 2 mg/mL, full symbols - ammonium chloride
6 mg/mL.

The synthesis of the proteinase in an exponentially growing culture is
controlled by amino acid repression, where complex mixtures of amino acids,
as well as a single amino acid, may be involved [7,8]. Quantitative
relationships between the amino acid(s) concentration and the specific rate
of proteinase synthesis ($r_E/\mu X$) were studied either during a 2 or 3 h
cultivation of a shaken culture supplemented with different amounts of
amino acid mixtures (Figure 3A) or during a short-term incubation with a
single amino acid (Figure 3B). In the first case a complex amino acid
hydrolysate (Oxoid casein hydrolysate) or a synthetic mixture of 9 amino
acids (aspartic acid, glutamic acid, alanine, leucine, isoleucine, valine,
serine, lysine and methionine) were added to the medium. The consumption
of the lowest amino acid concentration did not exceed 25% of the original
amount during the experiment.

The complete mixture of amino acids suppressed the proteinase synthesis by more than 90%. However, even in this case, some residual
activity was formed. The increasing amino acid concentration also
decreased to a certain extent the activity of the proteinase [9]. The
values in Figure 3A are therefore corrected for this inhibition. The
mixture of the 9 amino acids suppressed the enzyme synthesis less and its
non-repressible portion was larger. The relationship between the amino
acid concentration and the specific rate of the proteinase synthesis can be
expressed by the formula

$$r_E/\mu X = k_1 + k_2 \cdot e^{-k_3 \cdot S}$$

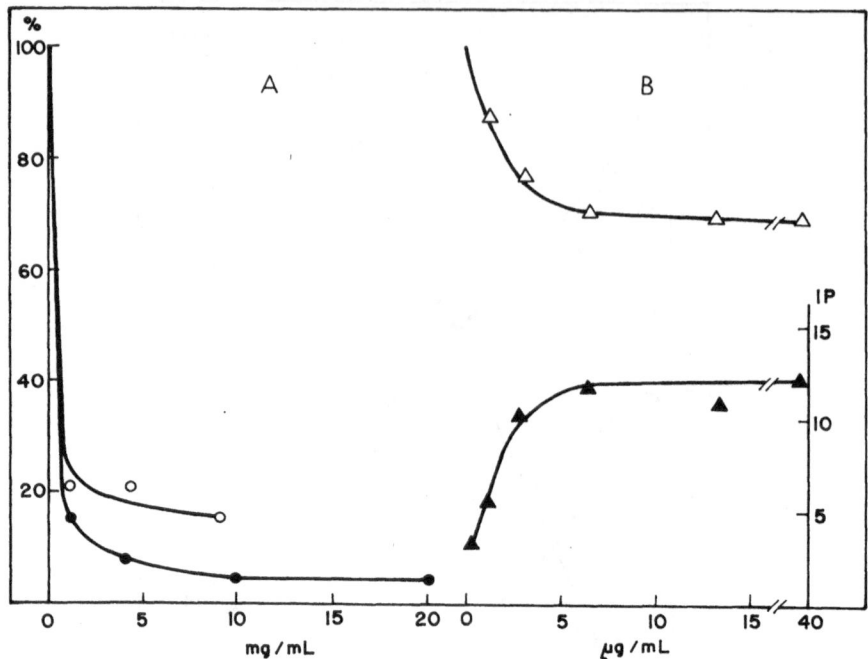

Fig. 3. Relationship between amino acid concentration and proteinase
repression. A - Abscissa: concentration of amino acid mixture
(mg/mL); ordinate: specific rate of proteinase synthesis in percentage of the control; full circles - casein hydrolyzate, empty
circles - mixture of 9 amino acids. B - Abscissa: concentration
of leucine (μg/mL); ordinate: empty triangles - specific rate of
enzyme synthesis, full triangles, IP - incorporation of ^{14}C-
leucine into cell proteins in dpm x 10^{-3} per 0.1 mg/mL dry matter.

where k_1 is the non-repressible portion of the proteinase, k_2 the repress-
ible portion, k_3 the repression coefficient and S the amino acid(s) concentration. The values of k_1 and k_3 were affected by the nature of the amino
acid mixture. The course of the curves in Figure 3A showed that even a
relatively low concentration of amino acids resulted in an almost maximal
repression of the enzyme. The amino acid concentration could not be
further reduced as it would decrease substantially during the experiment
due to the utilization of amino acids for protein synthesis. The proteolytic activity in the periplasm followed the activity in the medium, i.e.,
it was suppressed by amino acids to the same extent as was the excreted
enzyme. The activity in protoplasts, due to the presence of a serine
proteinase(s) was much less affected [9].

Other experiments were based on the "Ca^{2+}-pulse" technique; here the
incubation of the culture with Ca^{2+} proceeded for only 10 min and the
concentration of amino acid could be substantially reduced. The amino acid
mixture was replaced by a single amino acid (L-leucine), which did not
affect either growth or the activity of the proteinase and which suppressed
partially the enzyme synthesis. A decrease of leucine concentration in the
medium and the rate of its incorporation into proteins were determined
simultaneously. The results (Figure 3B) indicate that the saturation
concentrations of leucine for both the repression of the proteinase and for
the incorporation into proteins were approximately the same. It seems
therefore that the rate of the entry of the amino acid into the cell and/or

its entry into the amino acid pool used for protein synthesis may be crucial for the control of proteinase repression by amino acids.

Amino acids repress the proteinase formation at the level of mRNA transcription. This can be deduced from the experiment where either actinomycin D or amino acid mixture, both together with Ca^{2+}, were added to a population growing in its absence. The residual capacity coding for the enzyme decayed similarly in both cases (Figure 4). Half-life of the mRNA was about 5 min. The same half-life was also found for mRNA coding for the proteinase during sporogenesis [10].

Formation of the extracellular proteinase is also controlled by the cultivation temperature. Whereas the optimal growth temperature of B. megaterium is about 40°C or slightly more, the optimal temperature for the specific rate of the enzyme synthesis is more than 10°C lower (Figure 5). The suppressing effect of an increased temperature is not due to the inactivation of the proteinase, as the enzyme is relatively stable in the presence of Ca^{2+} at temperatures 40–42°C. The suppression of the proteinase formation by an increased temperature proceeds at the level of mRNA transcription and probably also during translation or excretion and could not be relieved by adaptation to elevated temperature. The temperatures of 40 – 42°C seem, however, to depress the excretion of all extracellular proteins [11].

REGULATION OF PROTEINASE SYNTHESIS DURING GROWTH AND SPOROGENESIS

Comparison of proteinase regulation during growth and development was based on the study of Bacillus megaterium 27, kindly provided by Prof. Udo Taubeneck from the Institute of Microbiology and Experimental Therapy, Jena, GDR. This strain produced less enzyme during growth than the asporo-genous strain KM but sporulated rather efficiently and synthesized the proteinase also during sporogenesis.

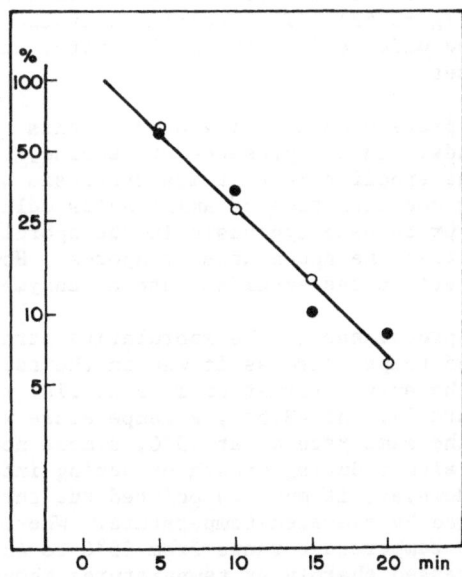

Fig. 4. Decay of the capacity to synthesize the proteinase. Empty circles – actinomycin D (0.75 μg/mL was added. Full circles – casein hydrolyzate (20 mg/mL) was added.

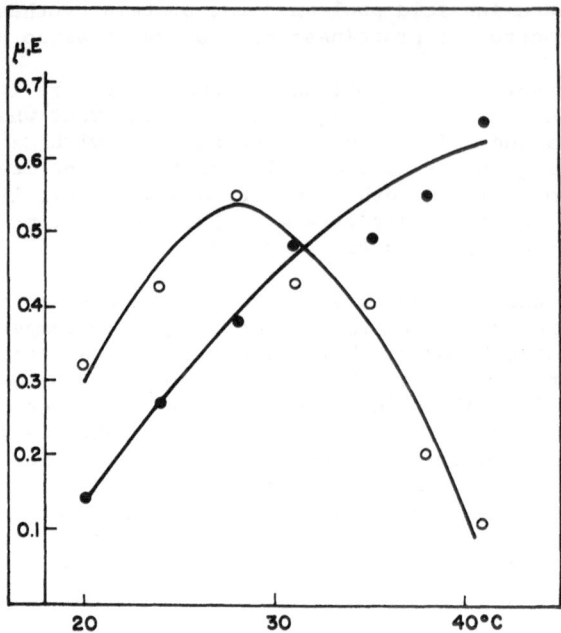

Fig. 5. Relationship between temperature and proteinase synthesis. Full circles, μ – specific growth rate; empty circles, E – specific rate of proteinase synthesis in TU/mg dry matter.

The enzyme was synthesized during the whole exponential phase and its activity in the medium increased exponentially, as did the cell dry matter (Figure 6A). However, its formation in sporulation medium was reduced to a limited time interval. It increased shortly after resuspension of the cells in the sporulation medium and ceased at the time when sporulation became irreversible (Figure 6B). The proteinase synthesis during sporulation thus seems to be under a developmental control and is repressed in later sporulation phases.

Formation of the proteinase during growth in this strain was also repressed by amino acids. In the presence of leucine, isoleucine and valine (3 mM each), the specific rate of its synthesis was half that in the control. The same concentration of amino acids delayed for about 1 hour the onset of the proteinase synthesis in the sporulation medium and postponed by the same time the appearance of spores. However, it was almost without any effect on the specific rate of enzyme synthesis [10].

Formation of the proteinase in the sporulating strain 27 was also suppressed by increased temperature as it was in the asporogenous KM. Temperature affected the enzyme formation in a similar way during growth and sporogenesis (Figure 7). At 43.5°C, a temperature which permitted growth to proceed at the same rate as at 35°C, almost no proteolytic activity was excreted either during growth or during incubation in the sporulation medium. However, it must be pointed out that sporulation itself was also impaired by elevated temperature. Whereas the growth rate was almost constant in temperature range from 35°C to 43.5°C, the sporulation efficiency decreased sharply at temperatures above 40°C (Figure 8A). The population incubated at 43.5-45°C in sporulation medium remained viable and capable of sporulating after having been transferred back to 35°C [12].

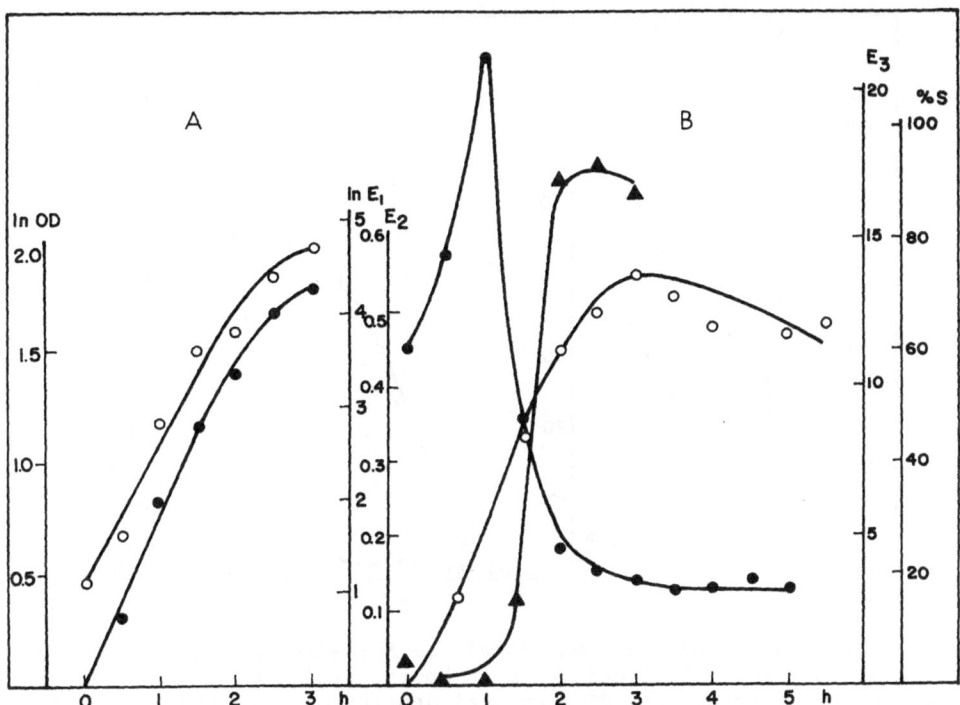

Fig. 6. Formation of proteinase during growth and sporulation at 35°C.
Abscissa - time in hours; A - growing culture: full circles; ln OD
- natural logarithm of optical density; empty circles, ln E1 -
natural logarithm of proteolytic activity (TU/L). B - sporulating
culture: empty circles, E2 - proteolytic activity (TU/10 mL), full
circles, E3 - specific rate of proteinase formation in units
(labelled substrate) per mg dry matter; full triangles - percentage of spores formed after addition of nutrients; control culture
was set equal to 100%.

The capability to form the proteinase after such a temperature shift-down
appeared after a short delay, indicating that its switching on was
inhibited by the increased temperature. This is in accordance with the
finding that the level of mRNA coding for the proteinase (as determined by
the "Ca^{2+} pulse - actinomycin D technique") was rather low in cells incubated at 43.5°C, as compared with the control population incubated at 35°C
(Figure 8B).

The inhibition of sporulation by netropsin, an antibiotic interacting
with AT groups in the chromosome and suppressing the synthesis of several
enzymes related to sporulation [13,14], caused an inhibition of protein
turnover but stimulated the rate of exocellular proteinase formation [15].
Netropsin not only increased the synthesis of the enzyme at the beginning
of incubation in sporulation medium but delayed also its late suppression
(Table 1). It must be stated, however, that the antibiotic stimulated the
proteinase formation also during growth and its effect thus does not seem
to be specific for sporulation [15].

The results did not provide unequivocal evidence whether the regulation of the proteinase during growth and sporulation are controlled by
different mechanisms or not. On the one hand, the switching off of the

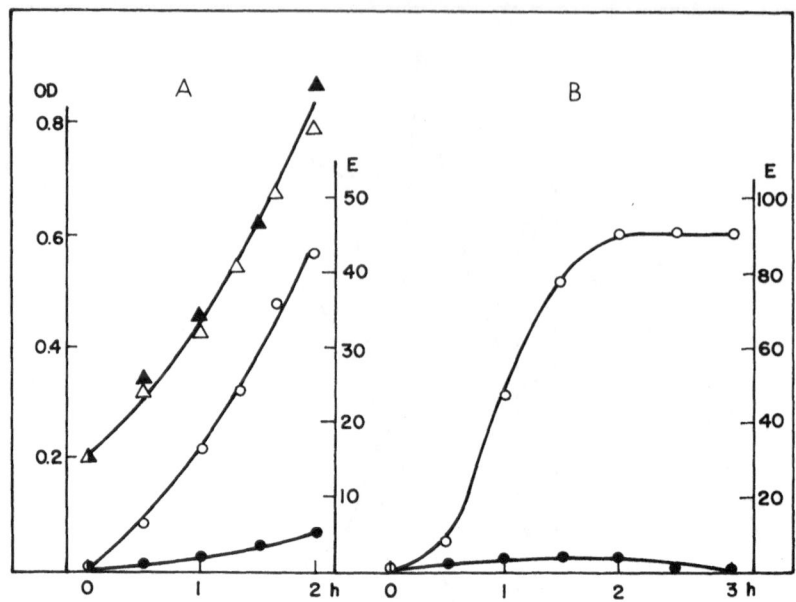

Fig. 7. Effect of temperature on proteinase formation during growth and sporulation. Abscissa: time in hours; ordinate: A – growing culture: triangles, OD – optical density of the culture; circles, E – proteolytic activity in units (labelled substrate) per mL; empty symbols – grown at 35°C, full symbols – grown at 43.5°C. B – sporulating culture: the symbols are the same; optical density 1.00.

enzyme synthesis at later sporulation stages as well as its relative resistance against amino acid repression favored the first possibility. On the other hand, the similar effect of netropsin and of elevated temperature on proteinase formation during growth and sporulation preferred the second explanation. However, the isolation of mutants differing from the wild strain in the formation of proteinase in growth or sporulation media supports the idea that the regulation of the enzyme under the two types of conditions is different [16].

The mutagenesis of germinating spores by ethyl methane-sulphonate resulted in isolation of several kinds of mutants (Table 2). Most interesting is the group C, to which belong mutants which synthesized less proteinase than the wild type during growth but more when incubated in sporulation medium. All mutants of that type were asporogenous. A typical mutant of this group, 27/36, was studied in more detail.

This mutant, together with the wild type, was grown in a synthetic medium with a reduced concentration of NH_4^+. The populations grew exponentially for about 2 h and the stationary phase was then brought about by exhaustion of the nitrogen source. The wild type formed refractile spores after 5-6 h of the stationary phase, the frequency of sporangia (60-80%) was comparable with that found in the sporulation medium. The specific rates of proteinase formation during exponential and stationary phases were determined by "Ca^{2+} pulses", the rate of protein synthesis by pulses of ^{14}C leucine. The specific rate of proteinase synthesis (expressed as units of enzyme per 10^3 dpm of incorporated ^{14}C leucine) in the wild type doubled at the beginning of the stationary phase and declined sharply afterwards (Figure 9A). This result is similar to that obtained in a sporulation

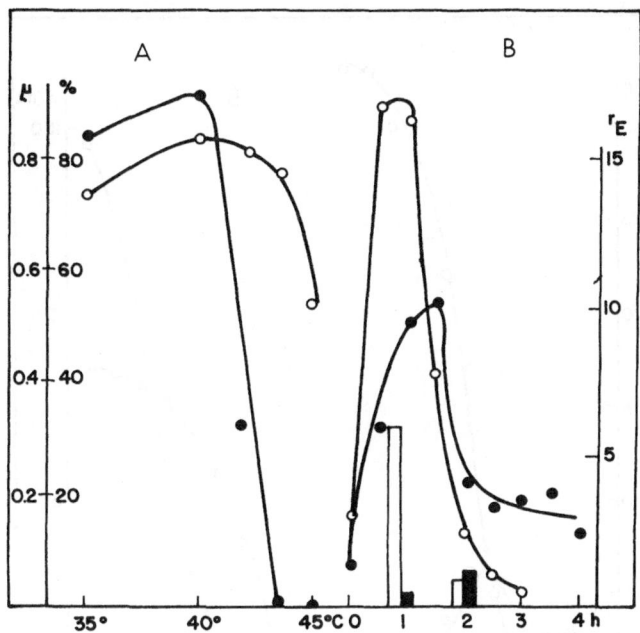

Fig. 8. Effect of temperature on proteinase formation during growth and sporulation. A – empty circles, μ – specific growth rate, full circles, %: percentage of spores formed after 6 h in the sporulation medium. B – circles, r_E – specific rate of enzyme formation; columns – capacity to form the proteinase in the presence of actinomycin D (1 μg/mL); empty symbols – incubation at 35°C, full symbols – incubation at 43.5°C.

Table 1. Effect of Netropsin on Specific Rate of Enzyme Synthesis in Sporulation Medium.

Hours	Control	0.5	1.0 μg netropsin/mL
1	4.0	5.2	7.9
2	2.1	6.2	11.1
3	0.7	4.7	4.6

Specific rate of enzyme synthesis is expressed in units (labelled substrate) synthesized per 10^4 dpm of ^{14}C-leucine incorporated.

medium. The mutant 27/36 formed the proteinase less efficiently during growth but the specific rate of enzyme synthesis increased sharply during the stationary phase. The total amount of enzyme excreted by the mutant to the medium exceeded that excreted by the wild strain less than would correspond to the differences in the specific rates of its synthesis (Figure 9B). This is evidently due to the decreased protein turnover in non-growing cells of the mutant, which was responsible for the lower rate of general protein synthesis.

CONCLUSION

Regulation of the extracellular proteinase synthesis during growth and during sporulation of B. megaterium has several common and several

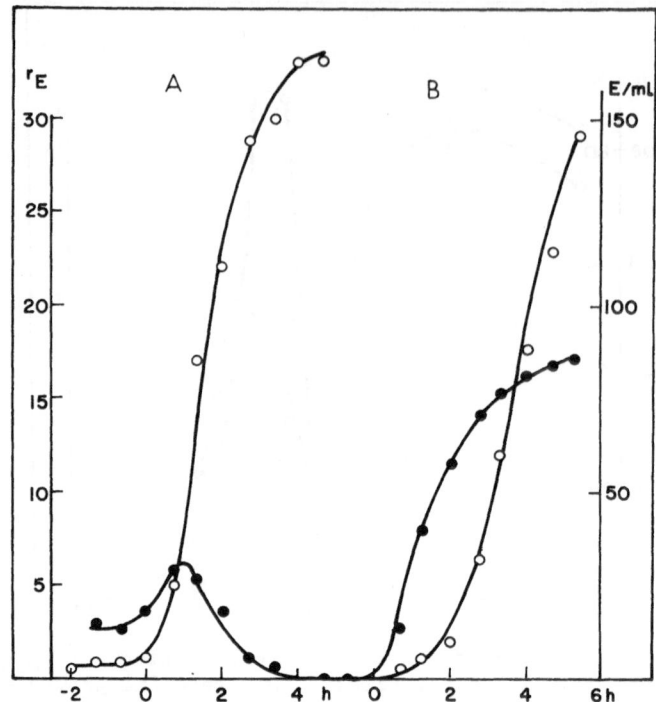

Fig. 9. Formation of proteinase in the wild type and asporogenous mutant 27/36. Abscissa - time in hours; A, r - specific rate of enzyme formation in units (labelled substrate) formed per 10^3 dpm of ^{14}C leucine incorporated; full circles - wild type, empty circles - mutant 27/36. B - E/mL, sum of enzyme excreted to 1 mL of the medium; the symbols are the same.

Table 2. Basic Characteristics of Proteolytic Mutants of B.megaterium 27.

| Group | Enzyme* | | Phenotype | Number of mutants |
	during growth	in sporulation medium		
A	1-2	3-5	spo	3
B	1	3-5	spo	6
C	0.1-0.5	2-4	spo	9
D	0.05	0.05-0.1	spo+	1

*Relative to the wild type whose enzyme production is taken to be unity.

different features. Whereas temperature and netropsin affected the enzyme formation in the same way during both physiological stages, amino acids repressed the proteinase only during growth. Moreover, asporogenic mutants were isolated that differ from the wild type in enzyme formation in growing and non-growing populations. They synthesized less proteinase during exponential growth and more during the stationary phase and seem to be impaired in switching off the enzyme formation, which is typical for sporulating cultures. These data suggest that B. megaterium either has two identical or very similar genes coding for the extracellular proteinase but differing partially in their promoter area or only one structural gene with a complex regulatory region. It is not possible at present to differentiate between these two possibilities.

REFERENCES

1. J. Mandelstam, Bacterial Sporulation: A Problem in the Biochemistry and Genetics of a Primitive Developmental System, Proc. R. Soc. London, Ser. B, 193:89 (1976).
2. F. G. Priest, Exocellular Enzyme Synthesis in the Genus Bacillus, Bacteriol. Rev., 41:711 (1977).
3. J. Millet and J-P. Aubert, Étude de la Megateriopeptidase, Protease Exocellulaire de Bacillus Megaterium. III. Biosynthèse et Role Physiologique, Ann. Inst. Pasteur, Paris, 117:461 (1969).
4. J. Millet, Étude de la Megateriopeptidase, Protease Exocellulaire de Bacillus Megaterium. I. Purification et Propriétes Generales, Bull. Soc. Chim. Biol., 51:61 (1969).
5. J. Chaloupka, M. Strnadová and V. Zalabák, Intracellular Proteolytic Activity during Sporulation of Bacillus Megaterium, Folia Microbiol., Prague, 22:1 (1977).
6. J. Millet, Caracterization d'une Endopeptidase Cytoplasmique chez Bacillus Megaterium en Voie de Sporulation, Compt. Rend. Acad. Sci., Paris, 272:1806 (1971).
7. J. Chaloupka and P. Krecková, Protease Repression in Bacillus Megaterium KM, Biochem. Biophys. Res. Commun., 8:120 (1962).
8. J. Chaloupka, P. Krecková and L. Rihová, Repression of Protease in Bacillus Megaterium by Single Amino Acid, Biochem. Biophys. Res. Commun., 12:380 (1963).
9. J. Moravcová and J. Chaloupka, Repression of the Synthesis of Exocellular and Intracellular Proteinases in Bacillus Megaterium, Folia Microbiol., Prague, 29:273 (1984).
10. J. Chaloupka, A. I. Severin, K. J. Sastry, H. Kučerová and M. Strnadová, Differences in the Regulation of Exocellular Proteinase Synthesis during Growth and Sporogenesis of Bacillus Megaterium, Can. J. Microbiol., 28:1214 (1982).
11. M. Vávrová and J. Chaloupka, Temperature Shiftup Suppresses Synthesis of Extracellular Proteins in Bacillus Megaterium, Current Microbiol., 12:9 (1985).
12. H. Kučerová and J. Chaloupka, Suppression by Temperature of Sporulation and of Exocellular Metalloproteinase Synthesis in Bacillus Megaterium, FEMS Microbiol. Letters, 28:293 (1985).
13. G. R. Keilman, B. Tanimoto and R. H. Doi, Selective Inhibition of Sporulation of Bacillus Subtilis by Netropsin, Biochem. Biophys. Res. Commun., 67:414 (1975).
14. B. L. Beaman, K. C. Burtis, R. H. Doi, J. P. Yoggy and D. P. Stahly, Ultrastructural Analysis of the Effect of Netropsin on Sporulation of Bacillus Subtilis, Can. J. Microbiol., 26:420 (1980).
15. H. Kučerová, M. Strnadová, V. Vinter, P. Graba and J. Chaloupka, Netropsin Stimulates the Formation of an Extracellular Proteinase and Suppresses Protein Turnover in Sporulating Bacillus Megaterium, FEMS Microbiol. Letters, 34:21 (1986).
16. H. Kučerová, L. Váchová and J. Chaloupka, Mutants of Bacillus Megaterium with Altered Synthesis of an Exocellular Proteinase, Folia Microbiol., Prague, 29:99 (1984).

STIMULATION OF A PROTEINASE SYNTHESIS IN BACILLUS

MEGATERIUM BY NETROPSIN

H. Kučerová and J. Chaloupka

Dept. of Enzyme Engineering
Institute of Microbiology
Czechoslovak Acad. Sci., Prague 4

Netropsin is a basic polypeptide antibiotic with cancerostatic properties having a high affinity for AT enriched sequences in the chromosome
[1]. Consequently, whereas growth of bacteria is only slightly affected,
sporogenesis is heavily impaired [2]. The process of sporulation seems to
be inhibited before or at the beginning of the II. sporulation stage [3].
We found that netropsin suppressed protein turnover but stimulated the
synthesis of an extracellular metalloproteinase in Bacillus megaterium [4].
The antibiotic affected in a similar way the formation of the proteinase
not only during sporulation but also during growth. The effect of the
antibiotic on the proteinase regulation was further studied and the results
are presented in this communication.

Netropsin (1 μg/mL) stimulated the formation of the proteinase both
during exponential growth and during an incubation in the sporulation
medium [5] (both at 35°C) after a one hour delay - Figure. 1. The synthesis of the enzyme was increased approximately two times in both cases.
The stimulation of the proteinase formation is due to an increase of the
synthesis of its m-RNA. This can be deduced from experiments, where Ca^{2+}
(0.5 mM) and actinomycin D (1 μg/mL) were added simultaneously to a culture
growing without Ca^{2+} and the residual capacity of the enzyme formation was
then determined. Since the proteinase is extremely unstable in the absence
of Ca^{2+}, the zero time proteolytic activity was negligible. The amount of
enzyme excreted during 3-min intervals was estimated for 30 min by a labelled protein substrate [6]; the theoretical value after 100 min of incubation was computed - Figure 2. In order to follow whether netropsin would
affect the formation or stability of mRNA coding for the proteinase,
netropsin was added either 100 min before actinomycin D or simultaneously
with this antibiotic. Netropsin increased the level of mRNA about 2-fold
(Figure 2A) but did not change its half-life, which was found to be about 5
min (Figure 2B).

The formation of the extracellular proteinase is repressed by amino
acids and by temperature [5,6]. The effect of netropsin on the formation
of the proteinase under repressive conditions was therefore examined.
Netropsin added together with amino acids could not stimulate the enzyme
formation as long as amino acids were present in effective concentrations.
No sooner than their amount decreased to a level permitting the formation
of the proteinase, a slight stimulatory effect of netropsin could be observed (Figure 3A). If the amino acids were supplemented one hour after

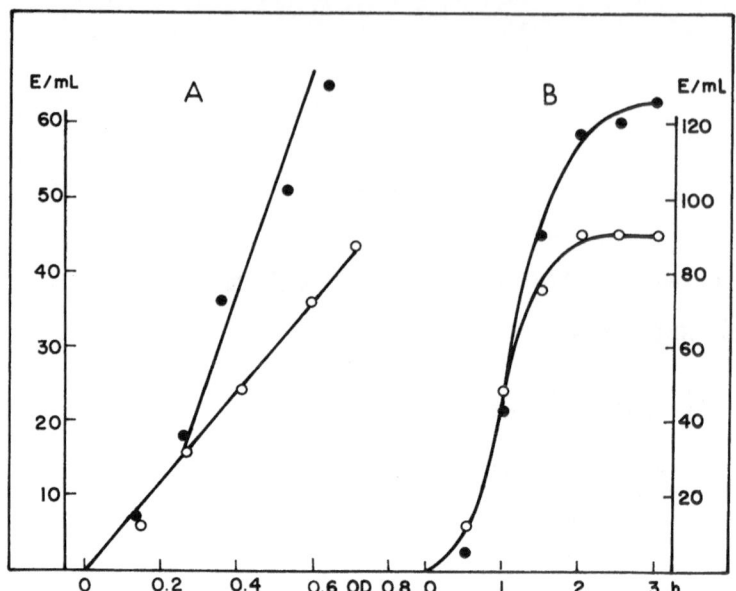

Fig. 1. Effect of netropsin on the formation of an exocellular proteinase during growth and sporulation. A – growth medium, B – sporulation medium; o – no netropsin, ● – with netropsin (1 µg/mL).

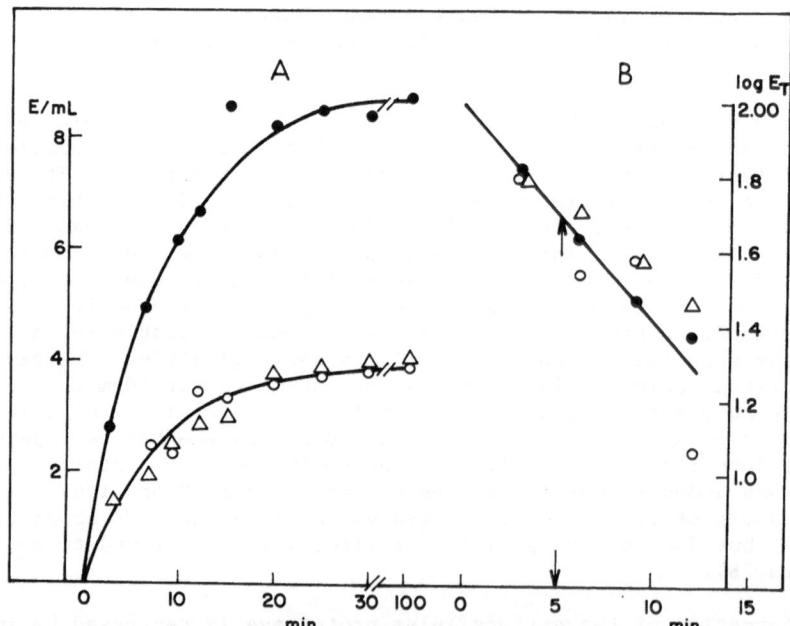

Fig. 2. Effect of netropsin on the synthesis and degradation of mRNA coding for the proteinase. Culture was grown for 100 min without Ca^{2+}, in the absence – o, △, or presence – ● of netropsin (1 µg/mL).
Actinomycin D (1 µg/mL) and Ca^{2+} (0.5 mM) were then added; – △ supplemented with netropsin after addition of actinomycin.

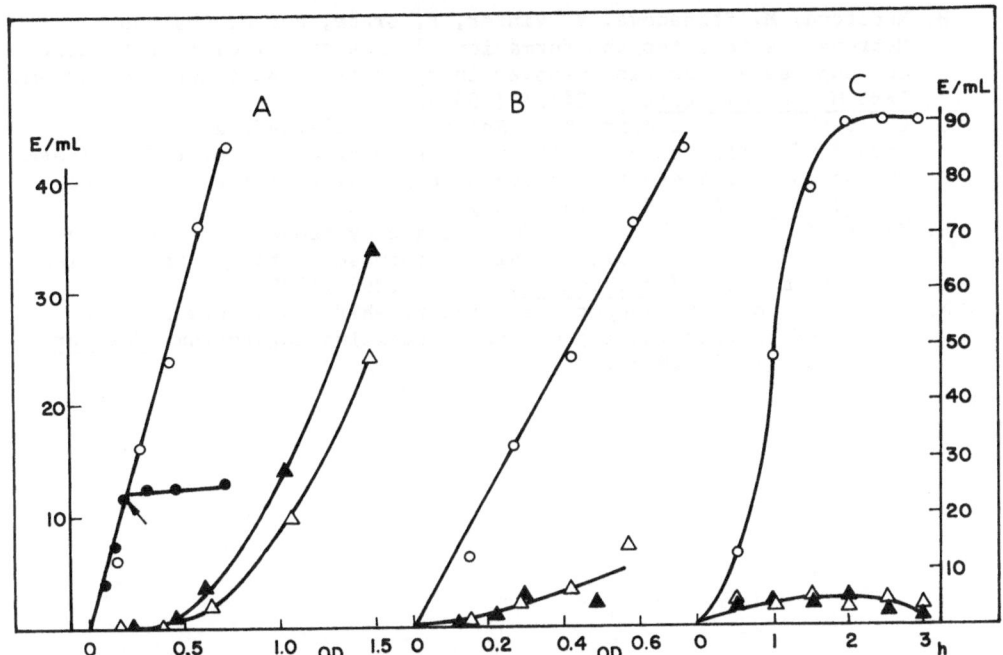

Fig. 3. Effect of netropsin on amino acid (A) and temperature (B, C) repression of the proteinase. A - o - control culture grown in the absence of netropsin and amino acids, ● - netropsin (1 μg/mL) added at zero time, casein hydrolyzate (2 mg/mL) 1 h later, △ - casein hydrolyzate added at zero time, ▲ - netropsin (1 μg/mL) added together with casein hydrolyzate at zero time. B - growth medium, C - sporulation medium; o - incubated at 35°C without netropsin, △, ▲ - cultivated at 43.5°C without (empty symbols) or with netropsin (full symbols).

netropsin addition, the enzyme synthesis was inhibited since the moment of their addition. Netropsin was also found not to relieve the repression brought about by a temperature shift up from 35°C to 43.5°C both during growth and during sporulation - Figures 3B, C.

The results give evidence that netropsin, while stimulating the proteinase formation under the conditions of derepression, is unable to relieve the repression of the enzyme caused either by amino acids or by elevated temperatures which are both known to affect the proteinase synthesis at the level of mRNA transcription [6,7].

REFERENCES

1. M. L. Kopka, C. Yoon, D. Goodsell, P. Pjure, and R.E. Dickerson, Binding of an antitumor drug to DNA. Netropsin and C-G-C-G-A-A-T-T-^{Br}C-G-C-G-, J. Mol. Biol. 183:553 (1985).
2. G. R. Keilman, B. Tanimoto, and R.H. Doi, Selective inhibition of sporulation of Bacillus subtilis by netropsin, Biochem. Biophys. Res. Commun. 67:414 (1975).
3. B. L. Beaman, K.C. Burtis, R.H. Doi, J.P. Yoggy, and D.P. Stahly, Ultrastructural analysis of the effect of netropsin on sporulation of Bacillus subtilis, Can. J. Microbiol. 26:420 (1980).

4. H. Kučerová, M. Strnadová, V. Vinter, P. Graba, and J. Chaloupka, Netropsin stimulates the formation of an extracellular proteinase and suppresses protein turnover in sporulating Bacillus megaterium, FEMS Microbiol. Letters 34:21 (1986).

5. J. Chaloupka, A.I. Severin, K.J. Sastry, H. Kučerová, and M. Strnadová, Differences in the regulation of exocellular proteinase synthesis during growth and sporogenesis of Bacillus megaterium, Can. J. Microbiol. 28:1214 (1982).

6. H. Kučerová, and J. Chaloupka, Suppression by temperature of sporulation and of exocellular metalloproteinase synthesis in Bacillus megaterium, FEMS Microbiol. Letters 28:293 (1985).

7. M. Vávrová, and J. Chaloupka, Temperature shift up suppresses synthesis of extracellular proteins in Bacillus megaterium, Current Microbiol. 12:9 (1985).

EXTRACELLULAR PROTEASE PRODUCTION BY B. AMYLOLIQUEFACIENS

M. J. Bawden, T. Litjens, T. R. Hercus,
B. K. May and W.H. Elliott

Department of Biochemistry, University of Adelaide
PO Box 498, Adelaide, SA, 5001, Australia

INTRODUCTION

B. amyloliquefaciens produces large amounts of two major secretory proteases: alkaline protease and neutral protease. Protease production by this organism is very unusual in its response to inhibitors of transcription. Using the antibiotics rifampicin and actinomycin D, which block transcription, we have demonstrated the presence of an apparent mRNA pool, capable of supporting protease production for about sixty minutes after addition of the antibiotics [1].

To examine this phenomenon at the molecular level required hybridization studies of protease mRNA production using a cloned gene probe.

ALKALINE PROTEASE mRNA PRODUCTION BY B. AMYLOLIQUEFACIENS

The gene for the alkaline extracellular protease of B. amyloliquefaciens was isolated using a cDNA probe synthesized using a synthetic oligonucleotide primer mixture specific for a short (14 nucleotide) segment of the gene (as derived from the protein sequence) and total mRNA isolated from the organism. The entire gene is contained within a 2.9 kb HindIII restriction fragment of DNA cloned in pBR322 (pBAP1).

This fragment has been completely sequences, and the gene sequence corresponds exactly with that of Wells et al., 1983 [2]. The protease is synthesized with a propeptide of as yet unclear function preceding the sequence of the mature protein. The cleavage site of the signal peptide has recently been determined by Vasantha and Thompson, 1986 [3].

The cloned alkaline protease gene was used as a probe to examine mRNA production by B. amyloliquefaciens, using the technique of Northern analysis.

Protease production is subject to repression by high levels of amino acids in the incubation medium. A build-up of alkaline protease mRNA occurs during incubation of cells in low amino acids medium (LAA) for various times, after incubation of these same cells in high amino acids medium (HAA) (repressive) for 75 minutes (Figure 1). Two transcripts are evident, the most prominent (1.4 kb) is of a size consistent with that

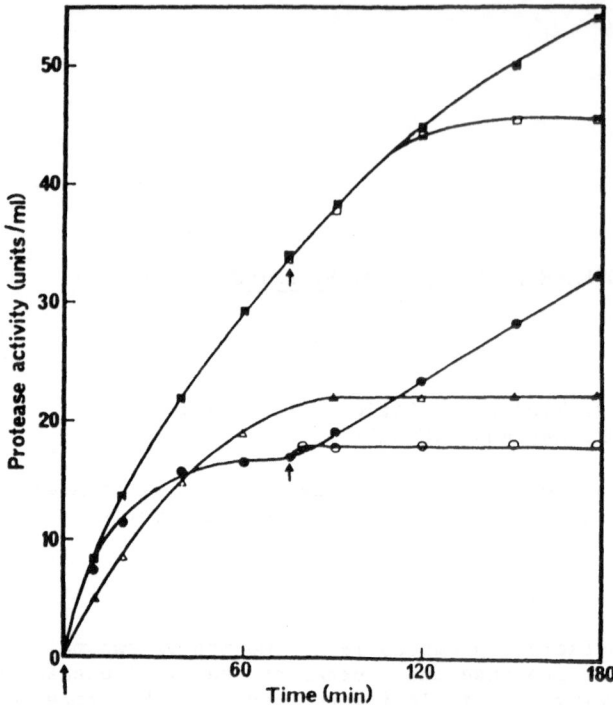

Fig. 1. Effect of rifampicin addition at zero and 75 minutes on protease production by washed cells of B. amyloliquefaciens in the presence of low and high amino acid levels. Cells were harvested at OD_{600} of 3,6, washed twice and resuspended to the same cell density in either low (0.025%) or high (0.5%) casamino acids medium and incubated with shaking at 30°C, rifampicin (0.5 µg/ml) was added at zero and 75 minutes to 20 ml samples of these cells. 1 ml samples were taken, centrifuged and the supernatants assayed for protease activity. (–■—■–) Low casamino acids, no addition of drug. (–●—●–) High casamino acids, no addition of drug. (–△—△–) High casamino acids, rimfapicin added at zero minutes, an identical curve was obtained in low casamino acids medium. (–○—○–) High casamino acids, rifampicin added at 75 minutes. (–□—□–) Low casamino acids, rifampicin added at 75 minutes. Arrows indicate time of rifampicin addition.

predicted from the DNA sequence of the cloned gene (Figure 2). The identity of the minor species (1.5 kb) is unclear, and remains to be investigated.

The decay of alkaline protease mRNA which occurs during prolonged incubation in low amino acids medium (LAA) was examined. Messenger RNA build-up was allowed to proceed for 30 minutes, at which time rifampicin was added to block transcription, and further time samples were taken. Using a scanning laser densitometer to measure band densities, a half-life of approximately 20 minutes was determined for the 1.4 kb transcript (Figure 3).

The findings presented here are consistent with the concept of a pool of unusually stable mRNA being responsible for the prolonged synthesis of alkaline protease in the presence of transcription inhibitors. The neutral protease needs to be examined in a similar manner.

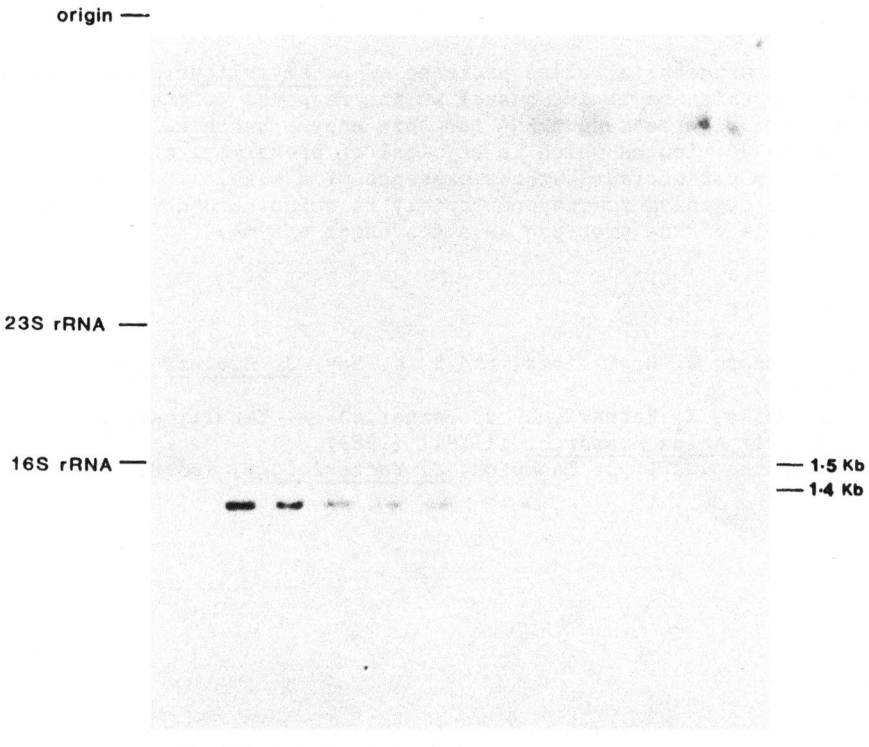

Fig. 2. Formation of the protease mRNA after the shift of repressed cells to a low amino acids medium.

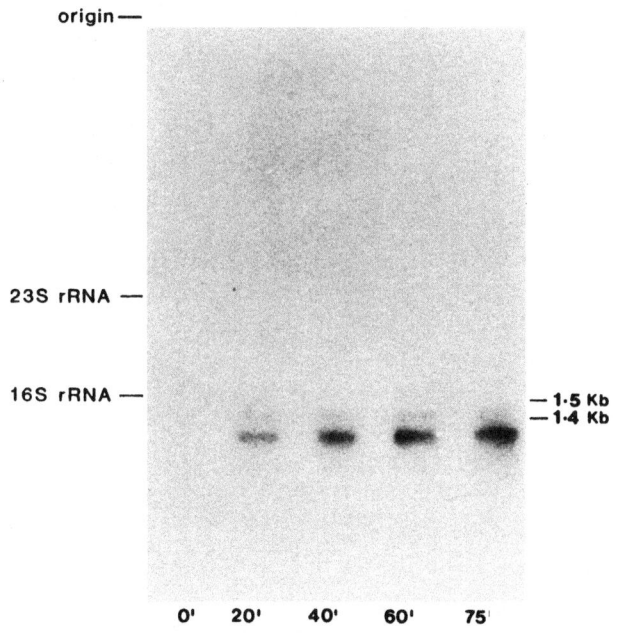

Fig. 3. Decay of the protease mRNA in the presence of rifampicin.

CONCLUSION

Our work with the alkaline protease of <u>B. amyloliquefaciens</u> shows that production of this enzyme is unusual in its response to inhibitors of transcription. The messenger RNA for this enzyme has a half-life of approximately 20 minutes which is atypical of prokaryote mRNAs. The reason for this is as yet unclear but the presence of a stable mRNA for protease may allow the organism to respond rapidly to amino acid deprivation, given the likely role of the protease as a scavenger enzyme.

REFERENCES

1. R. O'Connor, W. H. Elliott, and B. K. May, <u>J. Bacteriology</u>, 136:24 (1978).
2. J. A. Wells, E. Ferrari, D. J. Henner, D. A. Estell, and E. Y. Chen, <u>Nucleic Acids Research</u>, 11:7911 (1983).
3. N. Vasantha and L. D. Thompson, <u>J. Bacteriology</u>, 165:836 (1986).

FORMATION AND PROPERTIES OF A <u>BACILLUS</u> <u>SUBTILIS</u>

PROTEIN PROTEASE INHIBITOR

Arthur I. Aronson

Department of Biological Sciences
Purdue University, West Lafayette
Indiana 47907 USA

ABSTRACT

<u>Bacillus subtilis</u> protease inhibitor activity increases markedly at
the end of growth and during sporulation an increasing fraction of the
activity was released from cells. The inhibitor is unique in its amino
acid composition and the proteases affected. It is rather selective for a
major <u>B. subtilis</u> intracellular protease and could thus have a role in
modulating protease activity during sporulation but it was also active on
some fungal proteases. A substantial fraction was released from protoplasts implying a surface location. While there may be localization at or
near the membrane, the mechanism or site(s) probably differs from that of
alkaline phosphatase and penicillinase. The inhibitor protein gene has
been cloned in λgt11 and subcloned in a <u>B. subtilis</u> integration vehicle.
One subclone lacked inhibitor activity but still sporulated. A role in
modulating other intracellular processes or in cell protection are still
possible.

INTRODUCTION

A <u>B. subtilis</u> protein protease inhibitor active on a major intracellular serine protease was first described by Millet <u>et al</u>. [1,2]. The
inhibitor which could be released from cells by extended heating had a
molecular weight of 15-16,000 [2]. The increase of activity at the onset
of sporulation and its selectivity for a particular post-exponential
intracellular protease were consistent with a function in modulating
protease activity during sporulation.

Support for this possibility came from studies of Shimizu <u>et al</u>.
[3,4]. A Streptomyces inhibitor blocked sporulation of <u>B. subtilis</u> presumably by selective inhibition of a membrane bound protease. The latter was
postulated to degrade the <u>B. subtilis</u> protease inhibitor and thus
indirectly modulate the level of intracellular protease activity and consequently, sporulation. Burnett <u>et al</u>. [5] found that a substantial fraction
of the <u>B. subtilis</u> intracellular protease activity was cryptic i.e. antigen
was in excess of activity early in sporulation when protease inhibitor
activity was highest. This correlation was also consistent with a function
for the inhibitor in modulating intracellular protease.

While the inhibitor is selective for a particular B. subtilis protease, it is active on at least two fungal proteases [6]. In addition a substantial fraction of inhibitor activity is released upon treatment of cells with lysozyme. A surface location is thus possible, consistent with the presence of an E. coli inhibitor in the periplasm? The latter is active on subtilisin and pancreatic proteases but not on any of the known E. coli intracellular proteases. It should also be noted that the major intracellular serine protease of B. subtilis has been deleted by integration of a modified cloned gene with no adverse affect on sporulation [8]. There is thus no evidence that this protease is essential for sporulation except indirectly via a role in protein turnover [8].

Given the uncertainty in location and function of this protease inhibitor, further studies have been done on localization and possible function(s).

RESULTS AND DISCUSSION

As previously reported [6], a substantial fraction of protease inhibitor activity was released from protoplasts prepared from cells 1-3 hours after the end of exponential growth or by treating cells briefly (5-10 min) with lysozyme (20 μg/ml). At later stages of sporulation (T$_5$ was found in the low speed supernatant (Table 1). This loosely bound activity may account for the decrease in cell associated inhibitor activity i.e. phase white endospores), a larger fraction of the inhibitor activity reported previously [1,2]. The activity in the low speed supernatant was pelleted at higher speeds but 70% was retained in the soluble fraction if the low speed supernatant was incubated with lysozyme (20 μg/ml for 10 min at 25°C) prior to centrifugation 45,000 x g for 30 min. This result indicated binding of the inhibitor to cell wall material that was apparently fragmented (autolyzed) during the course of sporulation.

The protoplast and cell supernatants (Table 1) contained a major antigen of 15-16,000 that reacted with antibody to the purified inhibitor (unpublished results).

The inhibitor has been purified using a modification of the scheme of Millet and Gregoire [2] (Table 2) plus a final fractionation on a reverse phase HPLC column employing an n-butanol gradient (in 0.1% trifluoroacetic acid). The amino acid composition differs somewhat from inhibitors of E. coli or Streptomyces (Table 3) but they are probably all of the same general class in that they contain a low percentage of cysteine and are active on either subtilisin (E. coli and Streptomyces) or related serine proteases [8].

The release from protoplasts suggested a surface location perhaps analogous to that of some of the alkaline phosphatase [9,10] or penicillinase [11] activities in Bacilli. Some alkaline phosphatase in both the vegetative and sporulating cells is membrane bound, although the latter may be readily released by one or more proteases produced during sporulation [10] (Table 4). Extracts of sporulating cells (T$_3$) contained about 50% of the total alkaline phosphatase activity whereas similar extracts from exponentially growing cells contained less that 20% [10] (and unpublished results). A high salt wash removed most of the alkaline phosphatase activity [9] but little of the inhibitor activity (Table 4). Similar results were obtained with 1M MgSO$_4$. Either the localization or solubility properties of the inhibitor differ from those of alkaline phosphatase.

Much of the membrane bound penicillinase is anchored to the outside of the membrane via N-acyl glyceride cysteine although some activity is appar-

Table 1. Localization of B. subtilis Protease Inhibitor.

Cell Age	Fraction	Percent Total Inhibitor Activity[1]
T_{1-3}	Cell supernatant (9,000g/10 min)	<1
	Protoplast supernatant[2]	60-80
	Soluble extract[2]	15-30
	Insoluble fraction[2]	2-10
	1M MgSO$_4$ Wash[4]	<5
	1M NaCl Wash[4]	<5
T_5 (phase white endospores)	Cell supernatant (9,000g/10 min)	30[3]
	Protoplast supernatant	50
	Soluble extract	20
	Insoluble fraction	<5

[1] Assayed with B.subtilis intracellular protease and inhibitor units determined as previously described [6].

[2] Same results ± washes with 2mM PMSF, 1M NaCl and 0.03M Tris, pH 7.6.

[3] <10% if centrifuge cells at 18,000 x g/20 min or supernatant at 45,000 x g/30 min; 70% still soluble if treat 9,000 x g supernatant with lysozyme and then centrifuge at 45,000 x g/30 min.

[4] After the initial centrifugation of a 100 ml culture at 9000g for 10 min, the cell pellet was suspended in 2 ml, incubated at 25°C for 10 min and recentrifuged.

Table 2. Purification of B. subtilis Protein Protease Inhibitor.

	Total Activity[2] (x10^3)	Specific Activity[3] (x10^3)
1. Sonicate cells[1], heat supernatant 95°C/5 min and trichloroacetic acid precipitate between 5 and 10%	20	5
2. Dialyzed preparation in 0.02M Tris, pH 9.0, eluted from a DEAE-Sephadex G25 column with a linear gradient of 0.05-0.5M KCl in 0.02M Tris, pH 9.0	15	18
3. Peak from step 2, dialyzed, lyophilized, dissolved in distilled water and fractionated on an HPLC column[4]		
peak A	–	–
B	–	–
C	5	65
A+C	5	25

[1] From 12ℓ B. subtilis JH642 grown in NSM medium for 14 hr/37°C (30 min after the end of exponential growth)

[2] Inhibitory activity per ml

[3] Units per OD at 280 nm

[4] Gradient from 0 to 100% n-butanol in 0.1% trifluoroacetic acid. Peak C elutes at 25-26% n-butanol

Table 3. Amino Acid Content of Protease Inhibitors.

	B. subtilis	(moles %) E. coli[1]	Streptomyces[2]
Asp[3]	13.0	10.1	8.0
Thr	6.0	7.4	7.0
Ser	7.0	3.4	8.0
Glu[3]	14.0	18.2	5.0
Pro	4.0	6.1	7.0
Gly	11.0	7.4	10.0
Ala	10.0	6.8	16.0
(Cys)	2.2	1.4	4.0
Val	8.0	−	12.0
Met	ND	1.4	3.0
Ile	6.0	3.4	−
Leu	9.0	10.8	8.0
Tyr	0.4	8.1	3.0
Phe	3.0	2.0	3.0
His	5.0	0.7	2.0
Lys	5.0	11.5	2.0
Arg	2.0	−	4.0
Trp	ND	1.4	1.0

[1] Reference 7
[2] Ikenaka, T. (1985) Chapter 4 in Protein Protease Inhibitor – The case of Streptomyces Subtilisin inhibitor, pp. 159-163, (K. Hiromi, K. Akasaka, Y. Mitsui, B. Tonomura and S. Murao, eds.) Elsevier Science Publishers, Amsterdam
[3] Free acids plus glutamine or asparagine

Table 4. Relative Distribution of Protease Inhibitor and Alkaline Phosphatase.

	Percent of Total Activity or Units[1]	
	A'Pase	Inhibitor
5M NaCl wash	>90	12
Cell extract (2-140 kg)	<10	70
Pellet (P-140 kg)	−	18
French Pressure Cell (9,000 psi)		
Pellet$_1$ (8,000 g/10 min)	30	20
Pellet$_2$ (25,000 g/45 min)	20	22
Supernatant	50	58

[1] 500 ml of B. subtilis JH642 harvested two hours after the end of exponential growth. Alkaline phosphatase assayed as described elsewhere [9]. See footnote to Table 1 for washing protocol and determination of inhibitor units

ently removed from membranes by post-exponential proteases [11]. Such attachment may exist for the inhibitor, with release due to proteases activated by lysozyme treatment although extensive washing of cells with 1M NaCl and phenyl methyl sulfonyl chloride did not effect the percentage of inhibitor released by lysozyme.

The inhibitor gene has been cloned (Figure 1) by using the λgt11 expression system [12]. E. coli lysogens induced at 42°C with IPTG produced inhibitor active on the B. subtilis intracellular protease (the E. coli inhibitor is not active on this enzyme). The activity was released by treatment of cells with lysozyme-EDTA suggesting a periplasmic location (unpublished results). There was very little activity associated with the membrane fraction (12,000 x g for 30 min pellet of sonicated cells) suggesting no attachment via N-acyl glyceride cysteine.

An inactive EcoR1 fragment was subcloned into a shuttle vector, pDE194, capable of replication only in E. coli. In B. subtilis, the chloramphenicol-resistance gene was expressed upon integration of the plasmid at the site of homology i.e. the inhibitor gene (Figure 1). The partial diploids (selected at a low concentration of chloramphenicol in order to minimize amplification) should contain an intact and deleted inhibitor gene. Growth in the absence of chloramphenicol should result in loss of chloramphenicol resistance and one of the gene copies [8]. After plating, chloramphenicol-sensitive colonies were screened for sporulation and inhibitor activity. One isolate lacked inhibitor activity when the inhibitor fraction was prepared at several times during sporulation. This

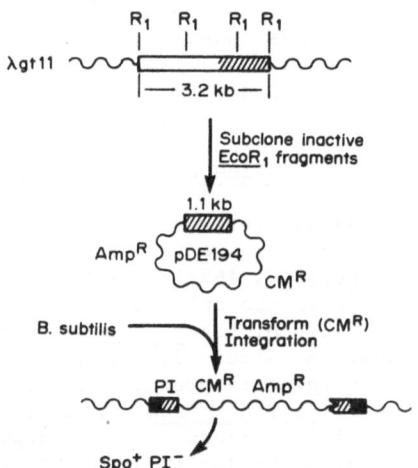

Fig. 1. Cloning protease inhibitor gene. Diagram of the construction of a B. subtilis strain lacking protease inhibitor activity. B. subtilis JH642 DNA was sheared in a French pressure cell at 7000 psi. Fragments were filled in, methylated with EcoR1 methylase, EcoR1 linkers attached and ligated into λgt11 arms [12]. Following packaging and infection of E. coli Y1088 to produce a phage lysate, E. coli Y1089 was lysogenized, plated and screened with inhibitor antibody (pre-adsorbed with E. coli Y1004 extract). After rescreening, a lysogen containing a 3.2 kb insert in λgt11 was isolated. Restriction enzyme digestion revealed two internal EcoR1 sites and a single Hind III site in about the center of the insert (not shown). A 1.7 kb Hind III-EcoR1 fragment (hatched area) subcloned in PUC 19 still produced inhibitor activity in E. coli. An EcoR1 digest of this subclone was ligated into the shuttle vector pDE194 capable of replication only in E. coli. Plasmid isolates were then transformed into B. subtilis JH642 selecting for chloramphenicol resistance. Following growth in liquid medium without chloramphenicol, cells were plated and individual colonies screened for sporulation and inhibitor activity after growth in 100 ml NSM.

isolate presumably contained a deleted gene from pDE194, an assumption that must be confirmed by Southern hybridization. There was no measurable alteration in the time of sporulation or the number of spores produced. The strain will be screened for other possible alterations in post-expontenial intracellular events i.e. protein turnover rates, amount of protease produced as well as possible susceptibility of sporulating cells to fungal proteases.

Acknowledgement

Research supported by grant GM20606 from the NIH. The technical assistance of Gail Sudlow is appreciated. Dr. J. Millet kindly provided unpublished results and rabbit antibody to purified inhibitor.

REFERENCES

1. J. Millet, Characterization of a protein inhibitor of intracellular protease from Bacillus subtilis, FEBS Lett., 74:59 (1977).
2. J. Millet and J. Gregoire, Characterization of an inhibitor of the intracellular protease from Bacillus subtilis, Biochimie, 61:385 (1979).
3. Y. Shimizu, T. Mishino, and S. Murao, Inhibition of sporulation of Bacillus subtilis by MAP1, a serine protease inhibitor, and inter-action of MAP1 with membrane bound protease, Agric. Biol. Chem., 48:365 (1984).
4. Y. Shimizu, T. Mishino, and S. Murao, Control of protease activity during sporulation of Bacillus subtilis, Agric. Biol. Chem., 48:3109 (1984).
5. J. J. Burnett, G. W. Shankweiler, and J. H. Hageman, Activation of intracellular serine proteinase in Bacillus subtilis cells during sporulation, J. Bacteriol., 165:139 (1986).
6. A. I. Aronson, Release from protoplasts of the Bacillus subtilis protease inhibitor, FEMS Microbiol. Lett., 33:47 (1986).
7. C. H. Chung, H. E. Iver, S. Almedand, and A. I. Goldberg, Purification from E. coli of a periplasmic protein that is a potent inhibitor of pancreatic proteases, J. Biol. Chem., 258:11032 (1983).
8. Y. Koide, A. Nakamura, T. Uozumiand, and T. Beppu, Cloning and sequencing of the major intracellular serine protease gene of Bacillus subtilis, J. Bacteriol., 167:110 (1986).
9. D. A. W. Wood and H. Tristram, Localization in the cell and extraction of alkaline phosphatase from Bacillus subtilis, J. Bacteriol., 104:1045 (1970).
10. J. A. Glynn, S. D. Schaffel, J. M. McNichols, and F. M. Hulett, Biochemical localization of the alkaline phosphatase of Bacillus licheniformis as a function of culture age, J. Bacteriol., 129:1010 (1977).
11. J. B. K. Nielsen and J. O. Lampen, Glyceride-cysteine lipoproteins and secretion by gram-positive bacteria, J. Bacteriol., 152:315 (1982).
12. R. A. Young and R. W. Davis, Efficient isolation of genes by using antibody probes, Proc. Natl. Acad. Sci. USA, 80:1194 (1983).

USE OF HIGHLY POROUS BEAD CELLULOSE WITH ATTACHED BACITRACIN

FOR AFFINITY CHROMATOGRAPHY OF A MICROBIAL PROTEINASE

J. Turková[1], M. J. Beneš[2], M. Kühn[3], V. M. Stepanov[4]
and L. A. Lyublinskaya[4]

[1]Institute of Organic Chemistry and Biochemistry,
Czechoslovak Academy of Sciences, 166 10 Prague,
Czechoslovakia
[2]Institute of Macromolecular Chemistry, Czechoslovak
Academy of Sciences, 162 06 Prague, Czechoslovakia
[3]Central Institute of Molecular Biology Academy of
Sciences of GDR, 1115 Berlin-Buch, GDR
[4]Institute of Genetics and Selection of Industrial
Microorganisms, Moscow, USSR

INTRODUCTION

Although affinity chromatography is currently used as a preparative
method on the laboratory scale, it is expected that its principle will be
utilized still more for large scale preparations of many proteins of
practical importance. For this purpose the requirements imposed on the
adsorbent, the method of its immobilization, and the carrier are very high.
A decisive role is played by the ligand. The cyclic nonapeptide bacitracin
(Figure 1), a noncompetitive inhibitor of many proteolytic enzymes, was
proved to be an efficient ligand for their purification by affinity
chromatography. It was attached to agarose or silichrom by cyanogen
bromide and benzoquinon, respectively (Stepanov et al., 1981). Both
carriers and their activation have some shortcomings which can be avoided
by using bead cellulose as a carrier and a different method of immobilization of the ligand.

Bead cellulose is a macroporous highly hydrophilic rigid material in
a regular spherical form (Štamberg et al., 1982; Štamberg and Peška, 1983).
It has excellent flow properties and a very low nonspecific adsorption.
Its volume practically does not change on variation of pH, ionic strength,
and in the presence of some organic solvents. It may be sterilized. It
cannot be used, of course, in the presence of cellulases. For the immobilization of bacitracin on the cellulose we used processes that give rise to
a stable bond of the ligand to the polysaccharide: activation of the cellulose with 2, 4, 6-trichlorotriazine or benzoquinone directly or by diazo-
tization of a 2-(4-aminophenylsulfonyl) ethyl derivative of cellulose.
Bacitracin-celluloses were used for the purification of subtilisin DY and
for isolation of alkaline proteinase from the culture medium of Bacillus
subtilis. The best results were obtained with a biospecific adsorbent
prepared by the attachment of bacitracin to bead cellulose activated by
2,4,6-trichlorotriazine.

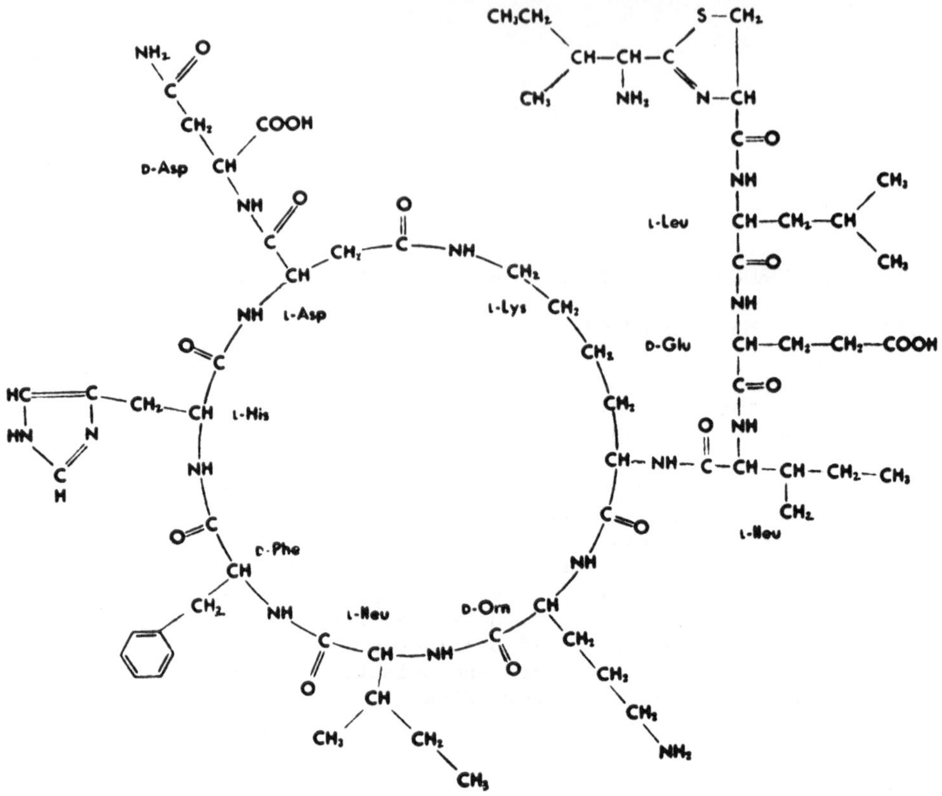

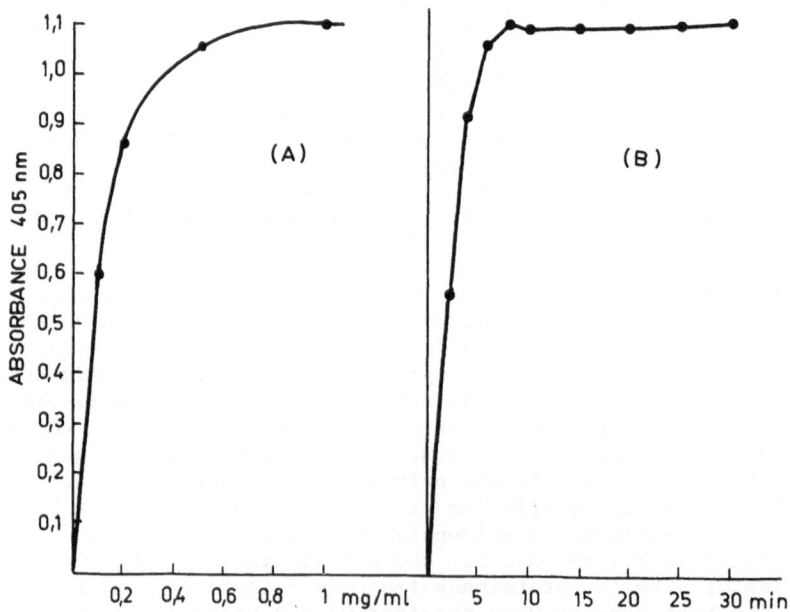

Fig. 1. Bacitracin A, $C_{66}H_{103}O_{16}N_{17}S$, white, hygroscopic amorphous powder, $[\alpha]_D^{23}$ +5° (±2.5°).

Fig. 2. Activity of subtilisin DY determined with benzyloxycarbonyl-L-alanyl-L-alanyl-l-leucine p-nitroanilide in dependence on the amount of the enzyme in mg/ml (A) and the time course of enzyme activity determination in min (B).

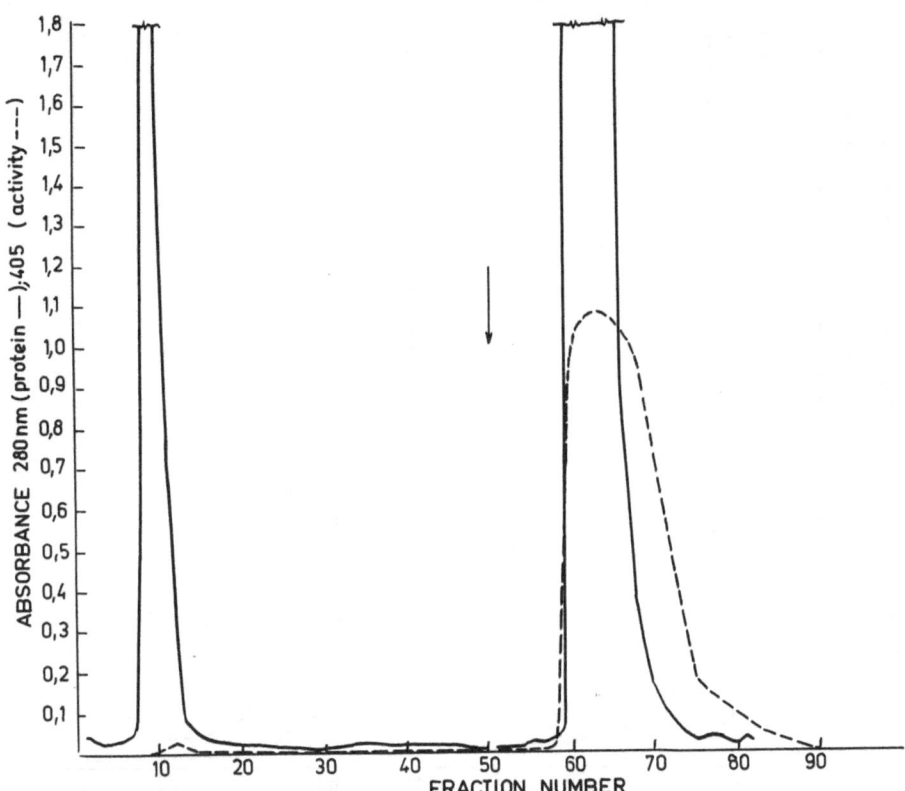

Fig. 3. Chromatography of subtilisin DY on bacitracin-cellulose (prepared
with 2,4,6-trichlorotriazine). 200 mg of subtilisin DY dissolved
in 3 ml of 50 mM Tris-HCl buffer, pH 8.3, containing 1 mM $CaCl_2$,
was applied on a column (30 x 2 cm) of bacitracin-cellulose
(inhibitor content 14.9 µmol/g of dry carrier) equilibrated with
50 mM Tris-HCl buffer, pH 8.3, containing 1 mM $CaCl_2$. After
washing the column with the equilibrium buffer, 20% isopropyl
alcohol in 1 M NaCl, buffered at pH 8.3 was applied at the
position marked with arrow. Fractions (6 ml) were taken at 5-min
intervals. Full line, protein; dashed line, activity determined
with benzyloxycarbonyl-L-alanyl-l-alanyl-L-leucine p-nitro-anilide
(Lyublinskaya et al., 1974; 1977).

Purification of subtilisin DY and isolation of alkaline proteinase
from culture medium of Bacillus subtilis.

Benzyloxycarbonyl-L-alanyl-L-alanyl-L-leucine p-nitroanilide is a
useful substrate for determination of the activity of subtilisin
(Lyublinskaya et al., 1974; 1977). Figure 2 (A) shows the activity of
subtilisin DY determined with this substrate as a function of the amount of
the enzyme and Figure 2 (B) shows subtilisin DY activity in dependence on
the time of enzyme activity determination.

Chromatography of subtilisin DY on bacitracin-cellulose prepared with
2,4,6-trichlorotriazine is shown in Figure 3. The yield of isolated subtilisin DY was 99.6%, the specific activity 5.1 unit/optical unit.

The same column and the same course of chromatography was very useful
also for the isolation of alkaline proteinase from culture medium of Bacil-

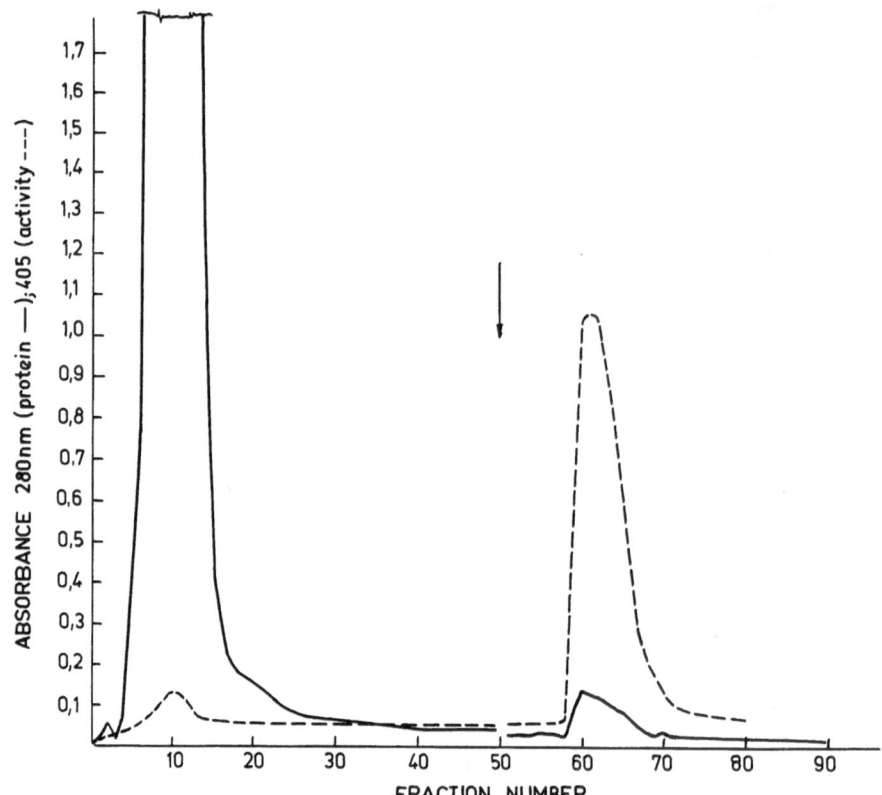

Fig. 4. Chromatography of culture medium of Bacillus subtilis on bacitracin-cellulose (prepared with 2,4,6-trichlorotriazine). The culture medium of Bacillus subtilis concentrated by ultrafiltration (10 ml) was applied on a column (30 x 2 cm) of bacitracin-cellulose. The procedure was identical with that shown in Figure 3.

lus subtilis (Figure 4). The yield of alkaline proteinase was 98%, the specific activity was 4.96 unit/optical unit.

CONCLUSION

Bacitracin-cellulose is an efficient affinity adsorbent for isolation of proteolytic enzymes. Its flow properties stability, and the possibility of a repeated use make it a promising adsorbent for large-scale operations.

REFERENCES

Lyublinskaya, L. A., Belyaev, S. V., Strongin, A. Ya., Matyash, L. F., Levin, E. D., and Stepanov, V. M. 1974, A new chromogenic substrate for subtilisin, Anal. Biochem., 62:371.

Lyublinskaya, L. A., Yakusheva, L. D., and Stepanov, V. M. 1977, Synthesis of subtilisin peptide substrates and their analogs, Bioorg. Khim., 3:273.

Štamberg, J., and Peška, J. 1983, Preparation of porous spherical cellulose, Reactive Polymers, 1:145.

Štamberg, J., Peška, J., Dautzenberg, H., and Philipp, B. 1982, Bead cellulose, in: "Analytical Chemistry Symposia Series, Vol. 9, Affinity Chromatography and Related Techniques", T.C.J. Gribnau, J. Visser and R.J.F. Nivard, eds., Elsevier, Amsterdam, p. 131.

Stepanov, V. M., Rudenskaya, G. N., Gaida, A. V., and Osterman, A. 1981, Affinity chromatography of proteolytic enzymes on silica-based biospecific sorbents, J. Biochem. Biophys. Methods, 5:177.

SACCHAROMYCES CEREVISIAE SECRETES X-PROLYL-DIPEPTIDYL (AMINO) PEPTIDASE: ELECTRON CYTOCHEMICAL STUDY WITH STATISTICAL EVALUATION

Josef Voříšek[1] and Kalju Vanatalu[2]

[1]Institute of Microbiology, Czechoslovak Academy of Sciences, Vídeňská 1083, CS 142 20 Praha 4 - Krc Czechoslovakia
[2]Institute of Chemical Physics and Biophysics, Estonian Academy of Sciences, SU 200 026 Tallin, Estonia SSR, USSR

The X-prolyl-dipeptidyl (amino) peptidase (DPP) activity was first detected by Suarez- Renduelles et al. (1981) in a particulate fraction isolated from haploid Saccharomyces cerevisiae. It was assayed with p-nitroanalide (pNA) derivatives of L-alanyl-L-proline (Ala-Pro) and Gly-Pro dipeptides as substrates. The amount of liberated pNA was measured spectrophotometrically. In intact cells the enzyme was accessible to the substrate after treatment with a nonionic detergent (Julius et al., 1983) and was concluded to be intracellular. An essential role is attributed to membrane-bound DPP in the final stages of proteolytic processing of the yeast mating pheromone (α-factor) precursor: the N-terminal sequence Glu-Ala-Glu-Ala is removed by DPP before the active pheromone is secreted from the cells. Genetic cloning distinguished two DPP activities: a heat-stable enzyme withstanding treatment at 60°C for 15 min and capable of splitting synthetic substrate Ala-Pro-pNA was called DPP A while 50-70% of total DPP activity that disappear after the heating was attributed to a thermo-sensitive DPP B. Both these activites were found in membrane fractions prepared from disintegrated yeast. Yeast mutant strains incapable of mating (ste) were proved to lack DPP A. It was concluded that DPP A and B were two different enzymes but their localization in the same or different membrane compartments was uncertain. The enzymes are probably not exocel-lular plasmalemma proteins. The export of DPP by membrane vesicles is more probable than its direct extrusion through the plasmalemma.

Bordallo et al. (1984) localized the thermosensitive DPP (B) in mem-branes of vacuoles liberated by isoosmotic lysis of yeast protoplasts. K_m for Ala-Pro-pNA was 0.06 mM. The heat-resistant DPP (A) (K_m 0.4 mM for Ala-Pro-pNA) was found in whole protoplast lysate. 80% of DPP B activity was in vacuolar membranes and 20% in the vacuolar supernatant.

According to the nomenclature of yeast proteinases (Achstetter et al. 1983) the particulate DPP A was called yscIV dipeptidyl (amino) peptidase and the particulate DPP B was called yscV DPP while the three newly reported soluble yeast DPP activities were called yscI to yscIII DPP.

The substrate specificity of purified thermo-resistant yscIV DPP was studied by the same team. The enzyme specifically hydolyzes peptide bonds involving the carboxyl group of prolyl residues penultimate to the unprotected termini, unless arginine is the N-terminal amino acid. However, the X-Ala-arylamide dipeptides included in the α factor precursor were not attacked. The authors concluded that the yscIV proteinase localized in the vacuolar membranes can hardly catalyze the processing of α factor that has to be secreted by membrane vesicles.

The electron-cytochemical method for localization of DPP was optimized by Lojda (1981) and its modification for cytochemical reaction for DPP in glutaraldehyde prefixed yeast cells was described recently (Vořišek in preparation). In principle, the synthetic dipeptide substrate has its carboxyl terminus bound to 4-methoxy-2-naphthylamine. The primary reaction product was formed by coupling hexazonium pararosaniline (HPR) with the MNA liberated by DPP and was osmiophilic. Thus postfixation by osmium tetroxide revealed also the electron-dense final reaction product on the background of the general osmiophilia of the matrix phase and membranes.

L-alanyl-L-proline-4-methoxy-2-naphtylamide (Ala-Pro-MNA) and Lys-Pro-MNA (BACHEM) were used as substrates in our study of DPP localization in an haploid strain of Saccharomyces cerevisiae X2180-1B and its pep4-3 mutant for (description of gene properties of the strains see Jones et al. 1982). The pep4-3 is a pleiotropic mutation yielding low levels of vacuolar proteinases A and B and of carboxypeptidase Y (10%, 7% and 3% of the levels in the wild parent strain, respectively) but normal X-prolyl-dipeptidyl (amino) peptidase activity (Suarez-Renduelles et al., 1981).

Results of cytochemical incubations are shown in Figure 1. At variance with the conventional electronograms of yeast fixed by glutaraldehyde and osmium tetroxide that feature a homogenous matrix phase and an electrondense vacuolar phase with lucent lines of all endoplasmic membranes, the cytochemical procedure yielded a transparent matrix phase with clearly depicted ribosomes and endoplasmic membranes of varying electron density. The background osmiophilia of the membranes without a DPP activity (the cytochemical control) is given here by the density of mitochondrial membranes (M). The DPP reaction product was distinct in the polar sheets of endoplasmic reticulum (ER) membranes and of the homologous membrane vesicles (V). However, the highest amount of DPP reaction product was pocked in small microcompartmens (circled) that were previously characterized as microglobules with an uniline boundary (therefore not vesicles) and radial spokes of coating fibres terminated by distinct granules (Vořišek, in preparation). The diameter of microglobules varied between 20-50 nm. Such coated microglobules fit well in to the newly postulated model of clathrin coated protein globules (nuggets without a membrane vesicle inside) that were found to prevail over the clathrin coated vesicles in the calf brain.

Not all of microglobules contained the DPP reaction product (double circled). In the wild haploid yeast the total number of microglobules with DPP activity was about one order of magnitude less than in the pep4-3 mutant.

The observed localization of DPP activity in the cytoplasmic microcompartments corresponds to the previously demonstrated function of this enzyme in the processing of the yeast mating pheromone (Julius et al., 1983) that is exocytosed and should therefore pass through the secretory path for yeast glycoproteins (for review see Schekman, 1982). However, the microglobules were not included in the above model where membrane vesicles represent the microconveyors. Of course, in any electronogram it cannot be unequivocally distinguished if the microconveyors (microglobules) move centrifugally (exo-cytosis) or centripetally (endocytosis) in the matrix phase. Their centrifugal translocation has been inferred from the cytochemical demonstration of glycoproteins in some of the microglobules and in the periplasmic space of yeast (Vořišek, in preparation). To this assump-

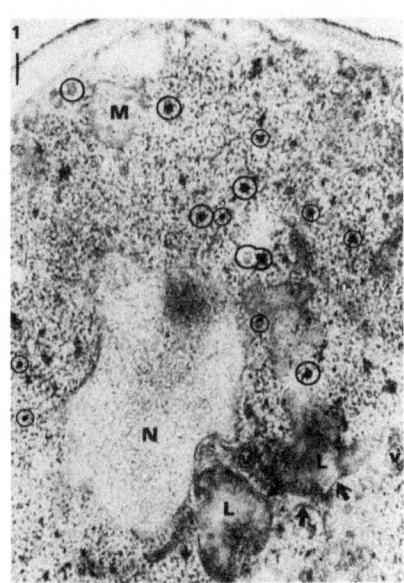

Fig. 1. Sectioned yeast <u>Saccharomyces cerevisiae pep4-3</u> mutant after
ultra-cytochemical reaction for lysyl-prolyl-dipeptidyl
(amino)-peptidase activity. L-lipoprotein, M - mitochondria,
N - nucleus, V - membrane vesicles; arrows indicate endoplasmic
reticulum: microglobules are circled. The bar marker indicates
100 nm.

tion corresponds well the observation of the DPP reaction product on the
periplasmic face of plasmalemma infoldings typical for yeasts (Figure 2).
Extensive deposits of the DPP reaction product were found also in the
periplasmic space underlying the yeast birth scars where new cell wall
constituents - glycoproteins - are deposited. As a rule, microglobules
with the DPP reaction product were situated near plasmalemma infoldings and
we conclude that at least the microglobules with DPP activity participate
in the exocytosis of yeast (glyco)proteins. At present the above ultra-
cytochemical results are the only indication that the DPP itself could be
exocytosed, probably together with the attacked protein. The previous
biochemical analyses were not sensitive enough for such conclusion.

The microglobules were distinguished in yeast for the first time and
number highly fluctuated among the cells of the mixed cell population
including all stages of the cell cycle. Also the periplasmic location of
DPP reaction product was observed in few cells. Therefore we found it
necessary to verify the significance of the above cytochemical observations
of cell sections by appropriate statistical calculations.

Let us call the microglobules (organelles) and the periplasmic deposits of the DPP reaction product cellular entities (CE). The probability
$\underline{p}$ of any CE to be inside an ultrathin section is $p = v / V$, when $\underline{v}$ is the
volume of the section and $\underline{V}$ the total cell volume. If there are $\underline{n}$ CE in
the cell then the probability $\underline{f}$ of finding x CE inside the section is

$$f_{(x)} = \binom{n}{x} p^x (1-p)^{n-x} \tag{1}$$

(formula for binominal distribution). The probability of finding any
number > 0 of CE inside the section is obtained by subtracting from the
total sum of probabilities ($\equiv$ 1 by definition) the probability for zero CE

107

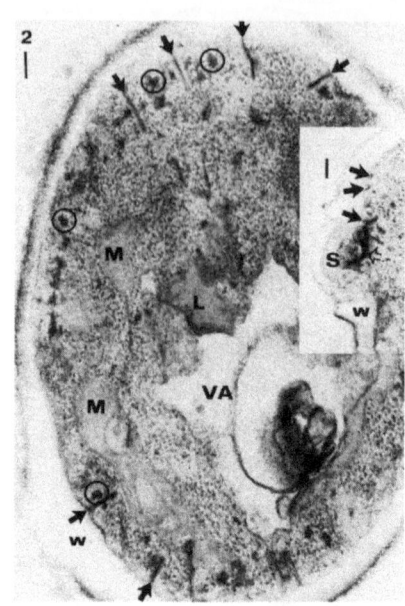

Fig. 2. Ultra-cytochemical reaction for lysyl-prolyl-dipeptidyl (amino) peptidase activity in a vacuolised yeast. L – lipoprotein; M – mitochondria; VA – vacuole; W – cell wall; microglobules are circled; arrows indicate invaginations of plasmalemma filled with the reaction product of proteinase. Insert shows the reaction product in the periplasmic space underlying the yeast birth scar (open arrow). Arrows indicate the plasmalemma invaginations; S – scar plug. The bar markers indicate 100 nm.

inside the section: $f = 1 - f_{(0)} = 1 - \binom{n}{0} p^{0} (1-p)^{n-0}$

assuming that $\qquad\qquad f = 1 - (1-p)^{n}$ (2)

(probability of finding any number of $\underline{n}$ CE inside the section, except zero). Let us assume we have $\underline{N}$ cells sectioned in one section plane. We use again the binominal distribution formula, only this time the probability $\underline{f}$ is the probability of a cell to show any CE ($\neq 0$) while sectioned. The probability of any of the $\underline{N}$ cells to "show" CE is:

$$F = 1 - \binom{N}{0} f^{0} (1-f)^{N-0} = 1 - (1-f)^{N}$$

or substituting $\underline{f}$ from (2) we get

$$F = 1 - (1-p)^{nN}$$ (3)

If we are watching an unsynchronized cell population and the CE is visible only during a period $\underline{t}$ of the total cell cycle time $\underline{T}$, then, instead of $\underline{N}$, we can actually observe the CE only in N. (t/T) cells and

$$F = 1 - (1-p)^{nN \frac{t}{T}}$$ (4)

Next we can take into account the dimensions of the CE relative to the section thickness. When the CE is large so that its diameter $\underline{d}$ is bigger than half of the section thickness h, then these organelles are discovered while being inside the plate with thickness h + 2d. In this case

$$p = \frac{v}{V} = \frac{(h+2d) \cdot \pi \cdot r^2}{V} \qquad (5)$$

where r is the cell cylinder radius. For the CE in question d = 20 to 50 nm and in this particular case h = 20 to 40 nm. To be more exact we can take into account the elliptical shape of a cross-section with radii a and b; then, instead of (5), we may write

$$p = \frac{v}{V} = \frac{(h+2d) \cdot \pi \cdot a \cdot b}{V} \qquad (6)$$

So we may use the following formula

$$F = 1 - \left(1 - \frac{(h+2d) \cdot \pi \cdot a \cdot b}{V}\right)^{nN \frac{t}{T}} \qquad (7)$$

For calculating V we assume that the shape of a yeast cell is a cylinder with diameter D and length L thus

$$V = \pi \cdot \left(\frac{D}{2}\right)^2 \cdot L \qquad (8)$$

We assume also that all the sections are perpendicular to the axis of a cylinder-shaped cell, hence in (6)

$$a \cdot b \equiv \left(\frac{D}{2}\right) \cdot \left(\frac{D}{2}\right) = \frac{D^2}{4} \qquad (9)$$

Now we write

$$F = 1 - \left(1 - \frac{(h+2d) \cdot \pi \cdot a \cdot b}{V}\right)^{nN \frac{t}{T}} = 1 - \left(1 - \frac{h+2d}{L}\right)^{nN \frac{t}{T}} \qquad (10)$$

For answering the question how many sections of individual cells (N') have to be analyzed in one section plane to be 99% certain to discover a CE, we consider F = 0.99 and calculate

$$\ln(1-F) = nN' \ln\left(1 - \frac{h+2d}{L}\right)$$

then

$$N' = \frac{1}{n} \frac{T}{t} \frac{\ln(1-F)}{\ln\left(1 - \frac{n+2d}{L}\right)} \qquad (11)$$

In numerical calculations, the actual parameters will be the following: L = 4000 nm; D = 1000 nm; d = 20 nm; h = 40nm; a CE is present n times in the cell while n = 1 or 10 or 100; we assume that the organelle is present during 1/3 of the cell cycle so that t/T = 0.33.

Numerical calculations were performed (on a microcomputer) for an ovoid cell (L = 4000 nm) sectioned randomly in the search for a CE of 20 nm diameter (a microglobule or a layer of DPP reaction product in the peri-plasm) that was present for 1/3 of the cell cycle period. The CE can be discovered, when alone in the cell, in one of 250 to 370 sections (variation is given by different D<L in the ovoid; for L = D ...., N' = 250);

when there are 10 CE in one cell - on one of 25 to 37 sections; with 100 CE present in one cell - in each third to fourth section.

The above calculations explained the observed high fluctuation of microglobule number among sections of different cells as a significant phenomenon relating to the transient character of secretory microconveyors. The appearance of a single microglobule in one section plane is then significant as well as the appearance of the DPP reaction product in a limited region of the periplasmic space of a yeast cell. In the opposite sense if we observe a CE in a single cell of the sectioned cell population and relate this observation to the number of analyzed cell sections, we can approximate the frequency of a CE in one cell of the population. As far as we know this is the first statistical evaluation of probability that a cell organelle will be observed on an ultrathin section of a cell from a mixed population and it can be concluded that the highly transient microconveyors participating in the secretory path of yeast can be found only after inspecting hundreds of sectioned cells. It was also possible to conclude that in the searching for DPP sites in yeast cells the ultra-cytochemical method was superior to the current methods of enzymological analysis of cell fractions and that the cytochemical approach extended the general knowledge of the secretory path. On the other hand, the cytochemical method could not distinguish between the thermostable yscIV DPP and the thermosensitive yscV DPP.

REFERENCES

Achstetter, T., Ehmann, C., and Wolf, D. H., 1983. Proteolysis in eucaryotic cells: aminopeptidases and dipeptidyltransferases of yeast revisited, Arch. Biochem. Biophys., 226:292.
Bordallo, C., Schwencke, J., and Suarez-Renduelles, M. P., 1984. Localization of the thermosensitive X-prolyl-dipeptidyl aminopeptidase in the vacuolar membrane of Saccharomyces cerevisiae, FEBS Lett., 173:199.
Jones, E. W., Zubenko, G. S., and Parker, R. R., 1982. PEP4 gene function is required for expression of several vacuolar hydrolases in Saccharomyces cerevisiae, Genetics, 102:665.
Julius, D., Blair, L., Brake, A., Sprague, G., and Thorner, J., 1983. Yeast α-factor is processed from a larger precursor polypeptide: the essential role of membrane-bound dipeptidyl aminopeptidase, Cell, 32:839.
Lojda, Z., 1981: Proteinase in pathology. Usefulness of histochemical methods, J. Histochem. Cytochem., 29:341.
Schekman, R., 1982. The secretory pathway in yeast, Trends Biochem. Sci., 7:243.
Suarez-Renduelles, M. P., Schwencke, J., Garcia Alvarez, N., and Gascon, S., 1981. A new X-prolyl-dipeptidyl aminopeptidase from yeast associated with a particulate fraction, FEBS Lett., 131:296.

PART III
AMYLASES

PRODUCTION OF α-AMYLASE BY BACILLUS STEAROTHERMOPHILUS
(pAT9) AND GENE MANIPULATION TO IMPROVE THE STABILITY
OF THE RECOMBINANT PLASMID

Shuichi Aiba, Yoshiaki Monden, Masatoshi Ohnishi,
Ming Zhang and Jun-ichi Koizumi

Department of Fermentation Technology
Faculty of Engineering, Osaka University
Yamada-oka, Suita-shi, Osaka 565, Japan

Abbreviations : Amy^+ presence of α-amylase gene; Amy^- absence of α-amylase gene; Km^r resistance to kanamycin; Km^s sensitivity to kanamycin; $PCase^+$ presence of penicillinase gene; $PCase^-$ absence of penicillinase gene; Sm^r resistance to streptomycin; Tc^r resistance to tetracycline

INTRODUCTION

Industrially important α-amylase genes from Bacillus amyloliquefaciens, Bacillus coagulans, Bacillus licheniformis and Bacillus stearothermophilus have already been cloned in vector plasmids pUB110 [24],
pBR322 [7], pBD64 [23] and in pTB90 [1], respectively, and expressed in
Bacillus subtilis [24], Escherichia coli [7], B. subtilis [23], and in B.
stearothermophilus [1]. It would be a matter of course that α-amylase
which is thermostable rather than thermolabile is more versatile. In
addition, α-amylase genes of barley [25], rats [18] as well as human beings
[22], etc. [21] have been transferred to E. coli by the use of either phage
or vector plasmid.

Apparently, the cloning of the α-amylase gene has been well documented. However, studies on cultivation of the transformant with a recombinant plasmid of the cloned gene are still scarce. However significant
the cloning per se might be from the aspect of strain breeding, it would be
of no less importance from the viewpoint of practice to study whether the
recombinant plasmid could be stably maintained during cultivation of the
transformant, and also, to search for a means of enhancing the stability
whenever needed.

It became clear from both batch and continuous cultures of B. stearothermophilus (pAT9) that the recombinant plasmid became unstable, yielding
a $Km^r Amy^-$ derivative especially when maltose instead of glucose was used
as carbon source [4], provided : that the recombinant plasmid, pAT9 (Km^r
Amy^+; 11.5 MDal (megadaltons)) is composed of a vector plasmid, pTB90 (Km^r
Tc^r; 6.7 MDal) and a HindIII chromosomal fragment (Amy^+; 4.8MDal) of B.
stearothermophilus [1]. It was also found that pLP11 (9.5 MDal; $Km^r Tc^r$
$PCase^+$) which was constructed by subcloning the penicillinase gene of
Bacillus licheniformis in the vector plasmid, pTB90 [8] became unstable
when the temperature of cultivation exceeded 50°C [2].

The purpose of this paper is to demonstrate (1) how the stability of pAT9 at 48-50°C could be enhanced by searching for a cause of instability in the level of DNA, and (2) how a vector plasmid pTRZ90 (Km^r; 7.9 MDal) that could grow in B. stearothermophilus even at 65°C was constructed from pTB90 by mutagenesis and DNA manipulation; the latter process gave rise to a recombinant plasmid pZAM26 (Km^r Amy$^+$, 8.7 MDal) that could produce α-amylase at temperatures above 50°C. Since the experimental techniques such as transformation [13], extraction of plasmid [10], use of restriction endonucleases to construct cleavage map [10], DNA sequencing [26], Southern hybridization [14], etc., all of which are indispensable for observations in both (1) and (2) have already been published separately [10,13,14,26], the description of these basic techniques used will be omitted here.

EXPERIMENTAL

Media and Chemicals

The composition of L broth is described elsewhere [16]. L broth supplemented with 0.25% (w/v throughout) of either maltose or glucose was denoted LM or LG broth, respectively. LM or LG broth that contained 0.05% K_2HPO_4, 0.05% KH_2PO_4, 0.08% $(NH_4)_2PO_4$ was designated LMP or LGP broth. When kanamycin [Km] 5 µg ml^{-1} was further added, it was called LMPK medium. L and LK agar, for instance were those media solidified by 2% of agar, respectively, while they were designated LS and LSK agar when supplemented with 1% soluble starch.

The companies and laboratories from which the antibiotics, restriction endonucleases, and all the reagents used here were purchased were the same as in the previous works [10,13,26].

Bacterial Strains, Plasmids and Phages

Bacterial strains, plasmids and phages used are listed in Table 1. B. stearothermophilus CU21 as well as glucose-repression-free GR2 [3] or GR201 were used as host in batch and/or continuous cultures. A recombinant plasmid pAT9 (11.5 MDal; Km^{r}Tc^r Amy$^+$) has a HindIII fragment (4.8 MDal), in which the α-amylase gene from B. stearothermophilus is cloned [1]. α-Amylase-deficient mutant of B. stearothermophilus AN174 [1] was used for the construction of a new recombinant plasmid pATHP9 (see later), while a vector plasmid pBR322 for subcloning a HindIII fragment of deletion plasmid (see later) to facilitate the establishment of restriction endonuclease map for the fragment. E. coli C600-1 [12] was the host of this subcloning, whereas phages M13mp10 and mp11 were used for DNA sequencing by dideoxy method, in which E. coli JM101 was the host [4].

Culture Conditions

Transformants of B. stearothermophilus CU21 or GR2 with plasmid pAT9 were precultured in LK broth at 48°C overnight for the seed. Cylindrical glass vessels (500ml working volume) with auxiliary devices that are described elsewhere [16] were used for continuous cultivation of the transformants at 48 to 50°C (inoculum size : 1%); shaken test tubes (5 ml) or a jar fermentor (1,000 ml, Marubishi, Type MD250, Tokyo) were used for batch cultures at 48 to 50°C. LMPK and/or LGPK media were used.

When a recombinant plasmid, pZAM26 (8.7 MDal; Km^{r}Amy$^+$) appearing later on in this work (not shown in Table 1) was examined for its performance at higher temperatures (53 to 62°C), the transformant of B. stearothermophilus GR201 was cultivated only in batch, using LGPK broth and the cylindrical vessels; the preculture was at 48°C overnight.

Table 1. Bacterial Strains, Plasmids and Phages.

Strain, plasmid or phage	Relevant characteristics	Reference
Bacillus stearothermophilus		
CU21	$Sm^r Amy^+$	13
GR2	$Sm^r Amy^+ Gcr^-$*	3
GR201	$Sm^r Amy^+ Gcr^-$	This work
AN174	$Sm^r Amy^-$	1
Escherichia coli		
C600-1	leu-6 thr-1 thi-1 supE44 lacY1 tonA21 hsdR hsdM Trp$^-$	12
JM101	Δlacpro thi supE F'traD36 proAB lacI^q ZΔM15	19
Plasmids and phages		
pAT9	$Km^r Amy^+$	1
pBR322	$Ap^r Tc^r$	6
M13mp10	lac	19
M13mp11	lac	19

* glucose catabolite-repression-free

Phenotypic examination of the transformant

Broth sampled from the culture was streaked on L agar plates, and they were incubated at 48°C for 24 to 48 h. One hundred colonies were transferred from single colonies on the plate onto LS and LSK agar plates by replica plating. These plates were incubated at 48°C for about 24 h to count the number of colonies having the Km^r or Km^s phenotype. The same plates were overlaid with an aqueous solution of KI-I$_2$ (0.01 M) to identify colonies exhibiting α-amylase overproduction (Amy^+) or absence of overproduction (Amy^-). The fraction of colonies with phenotype $Km^r Amy^+$ in the original one hundred colonies was defined as phenotypic stability of the recombinant plasmid.

Assays

(i) Cell concentration: OD_{660} reading of sample broth diluted three times was converted to dry cell weight per ml as described previously [2]. (ii) α-Amylase activity : α-Amylase activity was determined with the entire sample broth as saccharolytic power by the method of Bernfeld [5]. One unit was defined as the enzyme quantity that releases 1 mg of reducing sugar (as maltose) at 60°C for 3 min.

RESULTS AND DISCUSSION

As example of the continuous culture of B. stearothermophilus CU21 (pAT9) in LMPK medium at 48°C is shown in Table 2. The 2nd column, cycle number (D x t, where D : dilution rate, h^{-1}; t : time, h) corresponds to the time of broth sampling for the phenotypic examination. The appearance of deletion plasmid ($Km^r Amy^-$), deterioration of the plasmid (pAT9) stability, i.e. decrease of the phenotype ($Km^r Amy^+$) with the decrease of dilution rate, a total loss of the plasmid as represented by ($Km^s Amy^-$)

especially at D = 0.20 h^{-1}, and in addition, no exclusive loss of the
vector (pTB90) from pAT9 as manifested by (Km^{s}Amy$^+$) are apparent from the
table. This structural instability of pAT9 that is shown by (Km^{r}Amy$^-$) is
contradictory to the segregational instability (Km^{s}Amy$^-$) of pAT9 that was
observed in the previous work on LPGK medium at 47.5°C [3].

Whatever the cause of this contradiction might be, it is interesting
to see from Figure 1 that cultivation of the transformant yielded the same
size of the deletion plasmid irrespective of the difference in dilution
rate. It is also interesting and significant that the deletion plasmid
(Km^{r}Amy$^-$) appeared throughout in batch and continuous cultures of B.
stearothermophilus CU21 (or GR2) (pAT9), more frequently when maltose was
used, and that the molecular weight of deletion plasmids remained always
the same as in Figure 1. These deletion plasmids were designated pAT9sd
(spontaneous deletion) [4].

Restriction Endonuclease Cleavage Map of pAT9sd

Agarose gel electrophoresis of both plasmids, pAT9 and pAT9sd, cleaved
with HindIII showed that most of the fragment (4.8 MDal) encoding the
α-amylase gene in pAT9 was deleted and the fragment was reduced to about
0.6 MDal in pAT9sd. However, the rest fragments (vector plasmid pTB90; 6.7
MDal) of pAT9 and pAT9sd were kept intact. Accordingly, only the HindIII
fragment for pAT9 and pAT9sd are shown in Figure 2 to manifest the identity
and/or difference between these plasmids. Left and right terminals of the
HindIII fragment for pAT9 are designated HindIII (1) and (2).

Two DNA fragments from PstI to HindIII at both ends of the 4.8 MDal
cloned fragment in pAT9 were identical in size (about 0.9 MDal) and were
called A and B regions, respectively (Figure 2). Referring to a complete
nucleotide sequence of the α-amylase gene (open box in Figure 2; direction
of transcription indicated by an arrow) and its flanking regions [20], the
DNA fragment in A from the left end of the open box to HindIII (1) had two
TaqI sites ((2) and (1)) and four RsaI sites ((4) to (1)) (Figure 2).

The 0.6 MDal HindIII fragment of pAT9sd was subcloned at the HindIII
site of pBR322 in E. coli C600-1 and the plasmid DNA was extracted to
construct its cleavage map with TaqI and RsaI. The restriction sites of
these restriction endonucleases are shown in Figure 2. It is interesting
to find from the figure that cleavage maps of pAT9sd remained almost unchanged despite the wide difference in culture conditions.

RsaI (4) in pAT9 disappeared from the deletion plasmid, pAT9sd, whereas TaqI (3) that did not exist in pAT9 emerged in pAT9sd. However, the

Table 2. Phenotypic Behavior of Recombinant Plasmid pAT9 in a Continuous
Culture of the Transformant B. stearothermophilus CU21(pAT9);
culture medium, LMPK and 48°C.

Dilution rate D (h^{-1})	Cycle number D x t (−)	Phenotype (%)			
		Km^r Amy$^+$	Km^r Amy$^-$	Km^s Amy$^+$	Km^s Amy$^-$
0.20	2.9	18	72	0	10
0.39	3.0	54	46	0	0
0.61	3.3	72	28	0	0
0.81	3.1	62	37	0	1

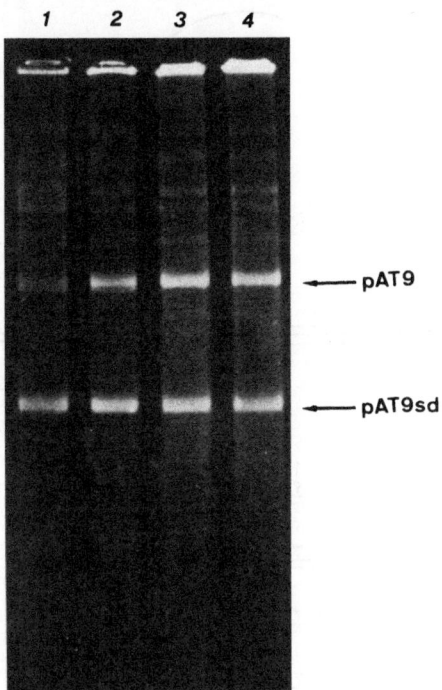

Fig. 1. Agarose gel electrophoresis of plasmids extracted from B.
stearothermophilus CU21(pAT9) in continuous cultures (LMPK broth)
at 48°C. Lanes 1 to 4: dilution rate, D = 0.20, 0.39, 0.61, and
0.81 h^{-1}, respectively.

resemblance of RsaI (and/or TaqI) cleavage sites between pAT9 and pAT9sd
suggested the need for DNA sequencing of the 0.6 MDal fragment of pAT9sd.
As a result of this sequencing [4], it was concluded that RsaI sites from
(1) to (3) on pAT9sd corresponded to those from (1) to (3) on pAT9. TaqI
(1) and (2) sites of pAT9 were reproduced in pAT9sd (Figure 2). Hence, the
left HindIII terminal of pAT9sd could be named HindIII (1). From the
finding that an area close to the left terminal of the pAT9sd HindIII
fragment came from the corresponding region adjacent to HindIII (1) of
pAT9, it was inferred that the deletion endpoint and junction (self-
ligation in pAT9sd) might be somewhere in a tiny fragment from RsaI (3) to
TaqI (3) of pAT9sd. This area had a stem/loop structure in pAT9 as well as
in pAT9sd [4].

Construction of Deletion Plasmids pATHP9 and pATP9202

pAT9 was subjected to double digestion with HindIII/PvuII, followed by
ligation with T4 ligase to construct a deletion plasmid that carries only
either A or B (Figure 2). The ligation mixture was used to transform B.
stearothermophilus AN174 by the protoplast procedure. A deletion plasmid
(designated pATHP9 ; 7.5 MDal) was extracted from transformed cells (Km^r
Amy$^+$) and identified (Figure 3, left). Likewise, another deletion plasmid
(named pATP9202; 8.3 MDal) was obtained from transformants (Km^{r}Amy$^-$)
(Figure 3, right).

Southern Hybridization

(1) ^{32}P-labeled DNA fragment (400 bp) from the right terminal of
HindIII to RsaI (3) in pAT9sd was used as probe. Double digest of pATP9202

117

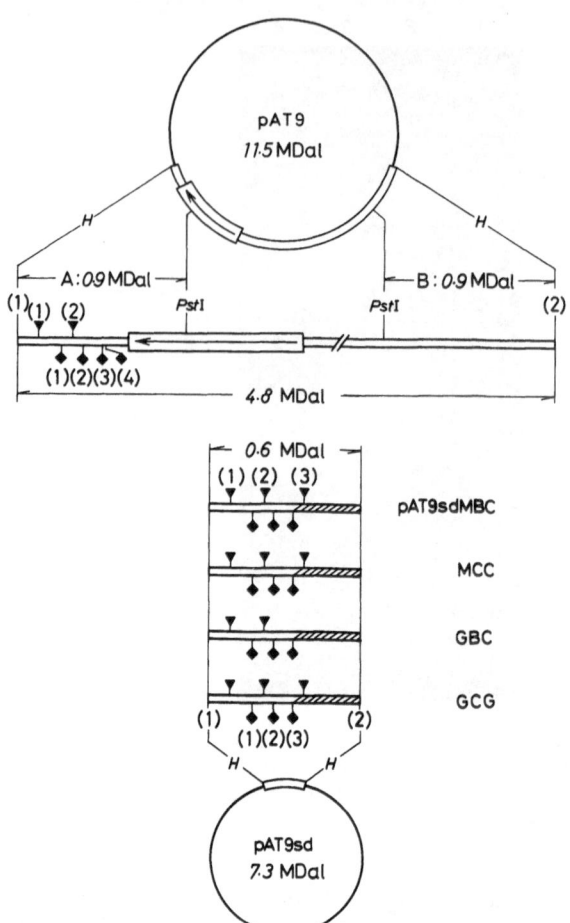

Fig. 2. Cleavage maps with restiction endonucleases for HindIII fragments
of pAT9 (top) and four pAT9sd series of plasmids. Symbols MBC,
MCC, etc. as the pAT9sd series of plasmids stand for the culture
conditions that gave rise to deletion plasmids, where 1st letter
denotes the carbon source, M: maltose, G: glucose; 2nd stands for
culture fashion, B: batch, C: continuous; 3rd for host strain, C:
CU21, G: GR2. Top bar represents pAT9 HindIII fragment (4.8
MDal), where an arrow inside the box (α-amylase gene) indicates
the direction of transcription[20]; both ends of HindIII sites (H)
are named (1) and (2), respectively; PstI is the cleavage site of
PstI; short vertical bars with solid triangles (▼) and diamonds
(◆) in the flanking region of α-amylase gene around the left end
of the pAT9 fragment are cleavage sites of TaqI and RsaI, respectively. These cleavage sites were deduced from the nucleotide
sequence of the flanking region [20]. RsaI sites, for instance,
were named (1) to (4) from left to right; HindIII to PstI fragments (0.9 MDal, each), A and B exhibited the sequence homology
(for details, see the text).
pAT9sd HindIII fragments are also shown. Each length (880 bp;
about 0.6 MDal) had cleavage sites of TaqI and RsaI that are shown
likewise by short vertical bars. These sites were also numbered
(1) to (3) from left to right. The hatched region around the
right terminal of the pAT9sd fragment represents the sequence
homology existing between this and B regions (for details, see the
text).

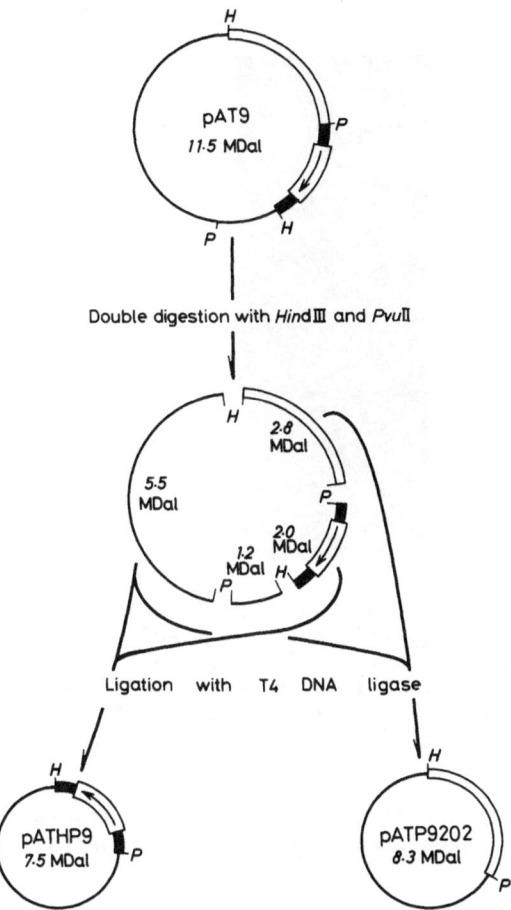

Fig. 3.   Construction scheme of deletion plasmids pATHP9 and pATP9202 from
          pAT9, and their cleavage maps with restriction endonucleases: H
          and P are HindIII and PvuII sites, respectively. Description of
          PstI site (Figure 2) is omitted here. For construction of
          pATP9202, single digestion of pAT9 with PvuI is sufficient. Open
          boxes are the α-amylase gene of B. stearothermophilus; the arrows
          within them are the direction of transcription.

(about 0.2 µg) with HindIII/PstI was electrophoresed on agarose gel (0.8%),
transferred to a nitrocellulose filter by Southern blotting and hybridized
to the probe (Figure 4, lanes 1 and 1', respectively).

     It is evident that the smaller fragment (corresponding to B in pAT9
(Figure 2)) could be hybridized to the probe. In other words, the right
terminal area of HindIII fragment in pAT9sd, whose DNA sequencing by the
dideoxy method failed, might have originated from somewhere in the vicinity
of HindIII (2). Therefore, the right HindIII terminal of pAT9sd could be
called HindIII (2).

     (2) ^{32}P-labeled DNA fragment B from HindIII (2) to PstI (0.9 MDal) of
pAT9 could be used as probe. Agarose gel electrophoreses of HindIII/PstI
double digests of pAT9, pATHP9 and pATP9202 (about 0.2 µg, each) are shown
in Figure 5 (lanes 1, 2 and 3, respectively). Although strong hybridiz-
ations in lanes 1' and 3' were self-evident owing to the use of the B
fragment as the probe, a weak hybridization in lane 2' was also observed.

119

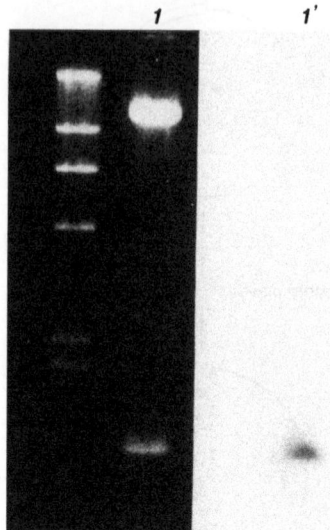

Fig. 4. Hybridization and blotting experiments. Double digest of plasmid
         pATP9202 with HindIII/PstI (see Figures 2 and 3) was electro-
         phoresed on argarose gel (lane 1; left marginal lane, HindIII
         digest of λ cI857 S7), transferred to a nitrocellulose membrane by
         Southern blotting, and hybridized to ^{32}P-labeled DNA fragment (400
         bp) from HindIII(2) to RsaI(3) of plasmid pAT9sd (lane 1').

These observations (lanes 2' (Figure 5) and 1' (Figure 4)) point out the
possibility that the area around HindIII (2) of pAT9sd came from both A and
B fragments of pAT9.

Stability of pATHP9

     Since the presence of the two homologous regions (A and B) around both
ends of the pAT9 HindIII fragment would have been deleterious to its
stability [15], it would be worthwhile to study here again the stability of
pATHP9 that was deprived of the B fragment from pAT9 (Figure 3).

     Transformants of B. stearothermophilus CU21 with pATHP9 were culti-
vated at 48°C in LMPK medium in batch and continuous fashions. Culture
conditions that would be most liable to cause deletions (Table 2) were
used, i.e. maltose as carbon source throughout and dilution rate at a
minimal level (=0.25 h^{-1}) in continuous culture. Another transformant, B.
stearothermophilus CU21 (pAT9) was re-examined as control. These results
are shown in Table 3. The enhanced stability and commendable performance
of plasmid pATHP9 are apparent from the table.

Construction of a Vector Plasmid pTRZ90 and Subcloning of α-amylase
Gene of B. stearothermophilus

     Although vector plasmid pTB90 isolated from the natural environment
[13] has been used to clone various genes (penicillinase [8], α-amylase [1]
and neutral protease [9]) in B. stearothermophilus, it was apparent from
cultivation of the transformant that the recombinant plasmid of penicillinase, for instance, became unstable, having lost either the cloned gene
or the entire plasmid when the cultivation temperature exceeded about 50°C
[2]. Whenever the recombinant plasmid becomes unstable, the transformant
cultivation per se turns out to be insignificant for evaluation of cloned

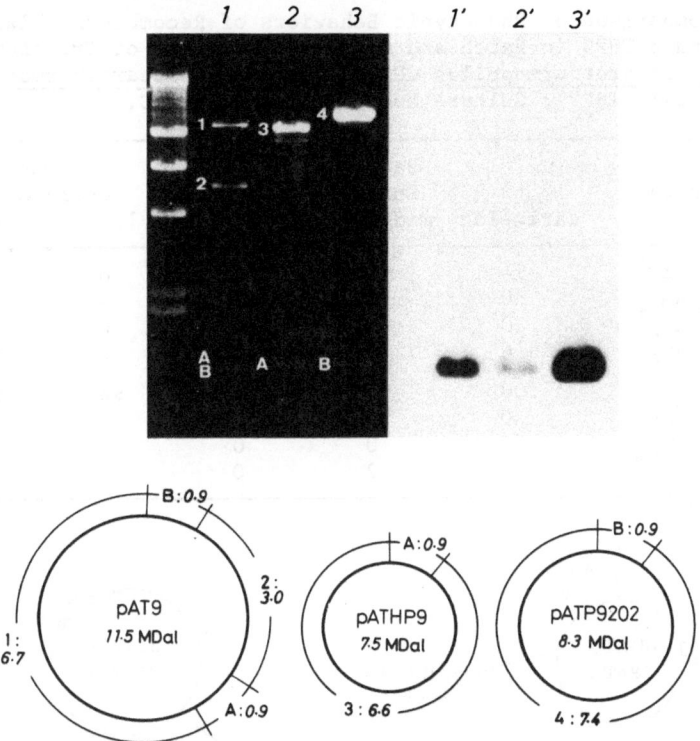

Fig. 5. Examination of homology between HindIII/PstI fragments A and B. Double digests (HindIII/PstI) of plasmids pAT9, pATHP9 and pATP9202 were electrophoresed on agarose gel (lanes 1 to 3, respectively; left marginal lane, HindIII digest of λ cI857 S7), transferred to a nitrocellulose membrane by Southern blotting, and hybridized to ^{32}P-labeled DNA fragment B (about 0.9 MDal) from HindIII(2) to PstI of plasmid pAT9 (lane 1' to 3'). A and B fragments in schematic diagrams below are the same as in Figure 2. DNA fragments numbered from 1 to 4 and also A and B fragments are marked on the agarose gel. The numeral after colon of each fragment designates its size in MDal. Fragment 1, vector plasmid, pTB90 (6.7 MDal); Fragment 2, chromosome DNA (about 3.0 MDal) of B. stearothermophilus; Fragment 3, DNA fragment (6.6 MDal) composed of (5.5 + 1.1) MDal coming from pTB90 and the chromosomal DNA (Figure 3), respectively; Fragment 4, DNA fragment (7.4 MDal) composed of (5.5 + 1.9) MDal coming from pTB90 and the chromosomal DNA (Figure 3), respectively.

gene expression, or even the transformant does not grow in the medium containing a drug as a selective pressure. Since the optimum growth temperature of B. stearothermophilus in LGP medium was from 60 to 64°C [2], it was desired to construct a vector plasmid such that the transformant could be stably cultivated at higher and/or up to the optimal growth temperature of B. stearothermophilus.

A schematic diagram of the construction of the plasmid pTRZ90 is shown on the left-hand side of Figure 6. A recombinant plasmid pLP11(9.5MDal), in which EcoRI fragment (2.8) MDal containing penicillinase genes from the constitutive-type of Bacillus licheniformis 9945A had been cloned by using pTB90 (6.7 MDal) in B. stearothermophilus CU21 [8] was the starting material for the construction.

Table 3. Comparison of Phenotypic Behaviors of Recombinant Plasmids pAT9 and pATHP9 in Batch and Continuous Culture of Transformants of B. stearothermophilus CU21(pAT9) and B. stearothermophilus CU21(pATHP9); Culture Medium, LMPK and 48°C.

Plasmid	Phenotype	Batch Phase			Continuous Cycle number (−)		
		early-log	mid-log	stationary	1.7	2.8	3.9
pAT9	Km^r Amy^+	94	96	90	88	74	47
	Km^r Amy^-	0	4	10	4	12	24
	Km^s Amy^+	0	0	0	0	0	0
	Km^s Amy^-	6	0	0	8	14	29
pATHP9	Km^r Amy^+	100	98	100	98	94	92
	Km^r Amy^-	0	0	0	0	0	0
	Km^s Amy^+	0	0	0	0	0	0
	Km^s Amy^-	0	2	0	2	6	8

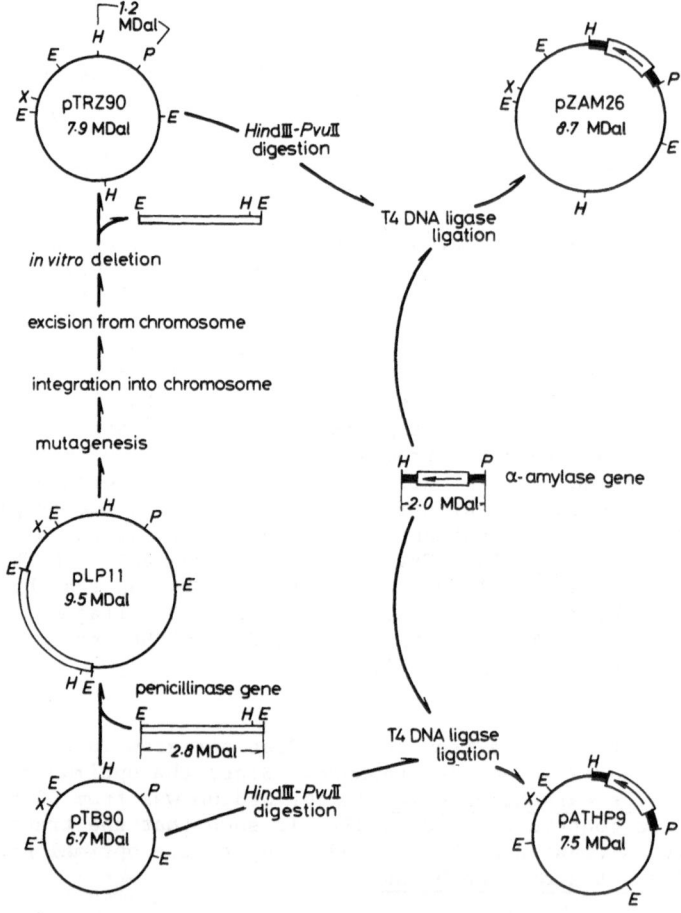

Fig. 6. Construction of the vector plasmid pTRZ90 and its subcloning of the α-amylase gene (recombinant plasmid, pZAM26). Open and solid bars are an EcoRI fragment containing penicillinase gene and a HindIII/PvuII fragment that contains the α-amylase gene (box with an arrow within). E, H, P, and X, are cleavage sites of EcoRI, HindIII, PvuII, and XbaI, respectively.

B. stearothermophilus carrying pLP11 was grown in LK medium (containing Km: 100μg ml $^{-1}$]; abbreviated hereafter at LKm$_{100}$) at 48°C to the midexponential phase and was harvested by centrifugation. The cells were suspended in L broth containing 10μg NTG (N-methyl-N'-nitro-N-nitrosoguanidine) ml^{-1} and 10 mM citric acid, and were incubated at 48°C for 15 min. The survival ratio after the mutagenic treatment was about 0.25. For the initial selection of mutants, the re-suspended bacteria were spread on LKm$_{500}$ agar and incubated at 60°C overnight. All colonies appearing on the first screening plates were transferred to L agar plates containing polyvinyl alcohol to test for penicillinase-producing (PCase$^+$) bacteria. Inocula from individual PCase$^+$ colonies were checked for growth in LKm$_{100}$ broth at 63°C (the third screening). B. stearothermophilus was transformed with plasmids extracted from colonies that passed the third screening.

The 45 strains obtained after the third screening were subjected to a repetition of the above screening procedures : only one of the strains passed. This candidate strain was cultured at 48°C and plasmid DNA was extracted. The plasmid (designated pTRA117) had the same size and cleavage map as pLP11 (9.5 MDal). B. stearothermophilus carrying either pLP11 or pTB90 could not grow in LKm$_{100}$ broth at 63°C, but when carrying pTRA117 it could (Table 4). The elevated upper growth-temperature of the transformant (63°C) was apparently due to the plasmid, because no specific treatments were imposed on the host strain other than the transformation procedure. Accordingly, pTRA117 was defined as a mutant plasmid of pLP11 [17].

It was found from the previous work [17] that pTRA117, existed as ccc DNA in the host of B. stearothermophilus when cultured in LKm$_{100}$ at 48°C turned out to integrate into the host chromosome when cultured at 63°C. It was also found previously [17] that when B. stearothermophilus (pTRA117) grown at 63°C in LKmh$_{100}$ broth was used to inoculate fresh LKm$_{100}$ medium at 48°C, a new plasmid (designated pTRZ117; 10.7 MDal) emerged as a result of excision of the former plasmid (pTRA117; 9.5 MDal) from the chromosome, carrying with it a 1.2 MDal DNA fragment of chromosome origin embedded near the repB region of pTRA117. Digestion of pTRZ117 with EcoRI, followed by ligation with T4 DNA ligase to deprive pTRZ117 of the 2.8 MDal fragment containing penicillinase genes yielded a vector plasmid pTRZ90 (7.9 MDal) as shown in Figure 6.

B. stearothermophilus carrying either pTRZ117 or pTRZ90 could grow in LKm$_{100}$ medium at higher temperatures (up to 65°C) than those carrying pTRA117 or pTB90 (Table 4). Double digestion of both pTRZ90 and pATHP9 with HindIII/PvuII, was followed by ligation with T4 DNA ligase, and the

Table 4. Growth of B. stearothermophilus Transformant in LKm$_{100}$ Medium.

| Plasmid | Temperature, °C | | | | | |
	48	55	59	63	65	67
pLP11	+	+	+	−	−	−
pTB90	+	+	+	−	−	−
pTRA117	+	+	+	+	−	−
pTRZ117	+	+	+	+	+	−
pTRZ90	+	+	+	+	+	−

Ability and inability to grow are shown by + and −, respectively.
+ implies an increase of OD$_{660}$ during several h to overnight culture, while − no increase of OD$_{660}$ after overnight culture. The host strain B. stearothermophilus CU21 could grow at 67°C in L broth without Km. Data for pTRZ919 were omitted.

ligation mixture was used to transform B. stearothermophilus AN174 (Sm^r-Amy$^-$) in order to subclone a 2.0 MDal HindIII/PvuII fragment containing α–amylase gene. The subcloning of the α–amylase gene in a recombinant plasmid pZAM26 (8.7 MDal) is shown on the right-hand side of Figure 6.

Performance of pZAM26 in the Production of α–amylase

A spontaneous mutant of B. stearothermophilus GR2 [3], designated GR201, was used as the host throughout. Both transformants, GR201 (pATHP9) and GR201 (pZAM26), were cultivated in batch using LGPKm$_{100}$ broth at 53 to 62°C, respectively. Cell concentration (mg dry cell ml^{-1}) and α–amylase activity of the broth (U ml^{-1}) were measured during each batch operation. The result is summarized in Figure 7.

The cell concentration is plotted against time (h), and the specific growth rate assessed is shown in each diagram near the specified period of time, at which the assessment was made. The specific growth rate of a transformant has been reported to deteriorate compared with that of the host cell. However, it is interesting to note from the diagrams at 53°C

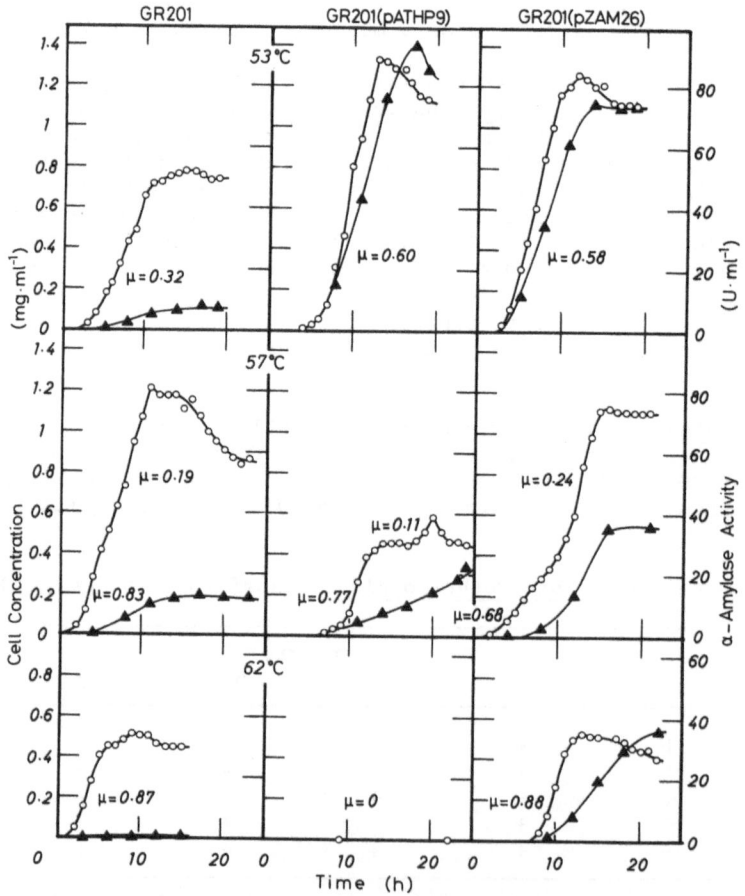

Fig. 7. Batch cultures of B. stearothermophilus GR201 and its trans-formants (pATHP9 and/or pZAM26) in LGPKm$_{100}$ broth at 53 to 62°C. μ–specific growth rate h^{-1} assessed at each specified period of time. Open (o) circles and solid (▲) triangles illustrate cell concentration (mg ml^{-1}) and α–amylase activity (U ml^{-1}).

that specific growth rate of the transformant was apparently larger than that of the host. The widely-accepted trend [11] of the transformant's growth rate was apparent only in the cultivation at 57°C.

It is evident in Figure 7 that the newly constructed vector plasmid, pTRZ90, which can be used at higher temperatures and its recombinant plasmid, pZAM26, having the α-amylase gene of B. stearothermophilus cloned, exhibited the expected performance that enhance the range of temperature beyond 50°C and close to the optimal growth temperature (62-64°C) of the host bacterium. Although not shown in the figure, the stability of pZAM26 in the host cells was in the range of 85-95°C throughout at 62°C.

CONCLUSION

1. Regardless of culture conditions of the transformant, B. stearothermophilus (pAT9 (11.5 MDal; Km^rAmy^+))(batch or continuous; glucose or maltose as carbon source), deletion plasmids (Km^rAmy^-) of the same size (7.3 MDal) emerged as the cultivation proceeded.

2. From DNA sequencing and Southern hybridization experiments on areas adjacent to both ends of the pAT9 HindIII fragment (4.8MDal) that contained the α-amylase gene, a dominant part (4.2MDal) of the HindIII fragment was confirmed to be lost, leaving tiny fragments near both terminals intact.

3. Because of the existence of homology between the fragments (0.9 MDal, each) contiguous to the pAT9 HindIII terminals, one of the fragments that did not encode the a-amylase gene was excluded and a recombinant plasmid pATHP9 (7.5 MDal; Km^rAmy^+) was constructed.

4. The phenotypic stability of pATHP9 in B. stearothermophilus was enhanced markedly in comparison to pAT9.

5. The upper growth temperature (59°C) of B. stearothermophilus (pTB90 (6.7 MDal, Km^rTc^r)) could be elevated to 65°C when pTB90 was replaced by a vector plasmid, pTRZ90 (7.9 MDal, Km^r.). The new vector plasmid was constructed by a series of steps including mutagenesis of pLP11 (9.5 MDal; $Km^rTc^rPCase^+$), its temperature-dependent integration in and excision from the chromosome of B. stearothermophilus.

6. A recombinant plasmid, pZAM26 (8.7 MDal; Km^rAmy^+) was constructed from pATHP9 (7.5 MDal; Km^rAmy^+) and pTRZ90 in B. stearothermophilus. The transformant, B. stearothermophilus (pZAM26), could produce α-amylase stably at 62°C.

7. In the cultivation of the transformant B. stearothermophilus (pAT9) to produce α-amylase, the phenotypic stability and culture temperature of the recombinant plasmid, pAT9 could be improved and elevated, respectively by the use of newly constructed recombinant plasmids, pATHP9 and pZAM26, in place of pAT9.

REFERENCES

1. S. Aiba, K. Kitai, and T. Imanaka, Cloning and Expression of Thermostable α-Amylase Gene from Bacillus stearothermophilus in Bacillus stearothermophilus and Bacillus subtilis, Appl. Environ. Microbiol, 46:1059 (1983).
2. S. Aiba, and J.-I. Koizumi, Effects of Temperature on Plasmid Stability and Penicillinase Productivity in a Transformant of Bacillus stearothermophilus, Biotechnol. Bioeng., 26:1026 (1984).

3. S. Aiba, J.-I. Koizumi, and J. S. Ru, Enhanced Production of α-amylase and plasmid stability in batch and/or continuous cultures of Bacillus stearothermophilus (pAT9), Chem. Eng. Commun., 45:217 (1986).

4. S. Aiba, Y. Monden, M. Ohnishi, J.-I. Koizumi, and M. Zhang, Unstable characteristics of a Recombinant Plasmid that has the Bacillus stearothermophilus α-Amylase Gene cloned and Its Stabilization in Bacillus stearothermophilus, J. Chem. Technol. Biotechnol. (in press).

5. P. Bernfeld, Amylases, α and β. Methods Enzymol, 1:149 (1955)

6. F. Bolivar, R. L. Rodriguez, P. J. Greene, M. C. Betlach, H. L. Heyneker, and H. W. Boyer, Construction and characterization of new cloning vehicles. II. A multipurpose cloning system, Gene 2:95 (1955).

7. P. Cornelis, C. Digneffe, and K. Willemot, Cloning and Expression of a Bacillus coagulans Amylase Gene in Escherichia coli, Mol. Gen. Genet., 186:507 (1982).

8. M. Fujii, T. Imanaka, and S. Aiba, Molecular Cloning and Expression of Penicillinase Genes from Bacillus licheniformis in the Thermophile Bacillus stearothermophilus, J. Gen. Microbiol., 128:2997 (1982).

9. M. Fujii, M. Takagi, T. Imanaka, and S. Aiba, Molecular Cloning of a Thermostable Neutral Protease Gene from Bacillus stearothermophilus in a Vector Plasmid and Its Expression in Bacillus stearothermophilus and Bacillus subtilis, J. Bacteriol.,154:831 (1983).

10. T. Imanaka, M. Fujii, and S. Aiba, Isolation and Characterization of Antibiotic Resistance Plasmids from Thermophilic Bacilli and Construction of Deletion Plasmids, J. Bacteriol., 146:1091 (1981).

11. T. Imanaka and S. Aiba, A perspective on the application of genetic engineering: stability of recombinant plasmid, Ann. N.Y. Acad. Sci., 369:1 (1981).

12. T. Imanaka, T. Tanaka, H. Tsunekawa, and S. Aiba, Cloning of the Genes of Penicillinase, penP and penI, of Bacillus licheniformis in Some Vector Plasmids and Their expression in Escherichia coli, Bacillus subtilis, and Bacillus licheniformis, J. Bacteriol., 147:776 (1987).

13. T. Imanaka, M. Fujii, I. Aramori, and S. Aiba, Transformation of Bacillus stearothermophilus with Plasmid DNA and Characterization of Shuttle Vector Plasmids Between Bacillus stearothermophilus and Bacillus subtilis, J. Bacteriol., 149:824 (1982).

14. T. Imanaka, T. Ano, M. Fujii, and S. Aiba, Two Replication Determinants of an Antibiotic-resistance Plasmid, pTB19, from a Thermophilic Bacillus, J. Gen. Microbiol., 130:1399 (1984).

15. I. M. Jones, S. B. Primrose, and S. D. Ehrlich, Recombination Between Short Direct Repeats in a RecA Host, Mol. Gen. Genet., 188:486 (1982).

16. J. -I. Koizumi, Y. Monden and S. Aiba, Effects of Temperature and Dilution Rate on the Copy Number of Recombinant Plasmid in Continuous Culture of Bacillus stearothermophilus (pLP11), Biotechnol. Bioeng., 27:721 (1985).

17. J. -I. Koizumi, M. Zhang., T. Imanaka and S. Aiba, Temperature-dependant Plasmid Integration into and Excision from the Chromosome of Bacillus stearothermophilus. J. Gen. Microbiol., 132:1951 (1986).

18. R. J. MacDonald, M. M. Crerar, W. F. Swain, R. L. Pictet, G. Thomas and W. J. Rutter, Structure of a family of rat amylase genes, Nature, 287:117 (1980).

19. J. Messing, New M13 Vectors for Cloning, Methods Enzymol., 101:20 (1983).

20. R. Nakajima, T. Imanaka and S. Aiba, Nucleotide Sequence of the Bacillus stearothermophilus α-Amylase Gene, J. Bacteriol., 163:401 (1985).

21. R. Nakajima, T. Imanaka and S. Aiba, Comparison of amino acid

sequences of eleven different α-amylases, Appl. Microbiol. Biotechnol., 23:355 (1986).

22. Y. Nakamura, M. Ogawa, T. Nishide, M. Emi, G. Kosaki, S. Himeno and K. Matsubara, Sequences of cDNAs for human salivary and pancreatic α-amylases, Gene 28:263 (1984).

23. S. A. Ortlepp, J. F. Ollington, and D. J. McConnell, Molecular cloning in Bacillus subtilis of a Bacillus licheniformis gene encoding a thermostable alpha amylase, Gene 23:267 (1983).

24. I. Palva, Molecular cloning of α-amylase gene from Bacillus amyloliquefaciens and its expression in B. subtilis, Gene 19:81 (1982).

25. J. C. Rogers and C. Milliman, Isolation and sequence analysis of a barley α-amylase cDNA clone, J. Biol. Chem., 258:8169 (1983).

26. M. Takagi, T. Imanaka and S. Aiba, Nucleotide Sequence and Promoter Region for the Neutral Protease Gene from Bacillus stearothermophilus, J. Bacteriol., 163:824 (1985).

THERMOSTABLE ALPHA AMYLASE OF <u>BACILLUS STEAROTHERMOPHILUS</u>:

CLONING, EXPRESSION, AND SECRETION BY <u>ESCHERICHIA COLI</u>

Ilari Suominen, Matti Karp, Jaana Lautamo*,
Jonathan Knowles* and Pekka Mantsälä

Department of Biochemistry, University of Turku
SF-20500 Turku, Finland
*VTT, Biotechnical Laboratory, SF-02150 Espoo, Finland

SUMMARY

The gene coding for a thermostable extracellular α-amylase was cloned from <u>Bacillus stearothermophilus</u> in <u>Escherichia coli</u> using lambda EMBL3 vector. The gene was subcloned in plasmids pUC-8 and pBR322. Nucleotide sequence of the gene was determined and it was shown to contain an open reading frame of 1650 bp coding for a preprotein of MW 63,000. The enzyme has a typical signal sequence for protein secretion. Processing of the signal peptide is ambiguous in <u>E. coli</u> occurring at two sites separated by three amino acid residues. The mature enzyme is homologous with two other liquefying α-amylases from <u>B. amyloliquefaciens</u> and <u>B. licheniformis</u>. The enzyme is initially secreted into the periplasm of <u>E. coli</u>, but when sufficient amount of α-amylase has accumulated in the stationary phase of growth, it is also found in the culture medium.

INTRODUCTION

All known species of the genus <u>Bacillus</u> produce exoenzymes and the variety of these enzymes is perhaps larger than in any other genus so far studied [1]. Being Gram-positive bacteria, these species secrete their exoenzymes into the growth medium. There are several thermophilic species in this group in which thermostability of both enzymes and cells can be studied. Many secreted enzymes of the genus <u>Bacillus</u> are also produced industrially and therefore are of commercial interest.

To study the structure-function relationships, particularly thermo-stability, of proteins by protein engineering and secretion of <u>Bacillus</u> proteins, we have begun studies of a thermostable α-amylase produced and secreted by <u>B. stearothermophilus</u>. We report here the cloning and the nucleotide sequence of the <u>B. stearothermophilus</u> α-amylase gene. We show that <u>E. coli</u> leaks mature α-amylase from the periplasm at the stationary phase. Ambiguous processing the signal peptide in <u>E. coli</u> is also reported. α-Amylases from other strains of <u>B. stearothermophilus</u> have been cloned previously [2,3,4] and while this paper was in preparation the nucleotide sequence of the gene was reported [5,6].

MATERIALS AND METHODS

In vitro packaging extracts were prepared from E. coli BHB2690 (recA, λ imm434, cIts, b2, red3, Dam15, Sam7/λ) and BHB2688 (recA, λ imm434, cIts, b2, red3, Eam4, Sam7/λ) [7]. E. coli NM538 (supF, hsdR) and NM539 (supF, hsdR, P2cox3) [8], were used as hosts for EMBL3 [8]. E. coli JM103 (Δ(lac-pro), thi, strA, supE, endA, sbcB15, hsdR4, F'(traD36, proAB, lacI²⁹ZΔMI5)) was used for subcloning in pUC-8 [9] and pBR322 and for sequencing experiments. M13mp10 and mp11 [10] were used for sequencing. The prototrophic B. stearothermophilus strain was from American Type Culture Collection (ATCC nr. 12980).

E. coli and B. stearothermophilus were both grown in LB medium [11] supplemented, when appropriate, with ampicillin (100 μg/ml) and/or starch (1%, w/v). For phage lambda, LB medium was supplemented with 10 mM MgSO₄ and 0.4% maltose (LAM). For growing phage M13, E. coli JM103 was grown in 2xYT liquid medium or on B-plates [10].

Total cellular DNA from B. stearothermophilus and plasmid DNA were isolated by previously described methods [11,12] and purified by CsCl gradient ultracentrifugation or by Sephacryl S-1000 chromatography [13]. Phages and phage DNAs were isolated and purified by standard methods [10,11].

Restriction enzyme digestions and DNA ligations were performed with enzymes from commercial sources by standard procedures [11].

For sequencing, plasmids (see RESULTS) were cleaved (single or double digestions) with various restriction enzymes and appropriate fragments cloned in M13mp10 and/or mp11. The fragments were sequenced with the dideoxy method [14] using ³⁵S-dATP (Amersham, 600 Ci/mmol) [15].

α-Amylase activity was measured by following the increase in reducing capacity [16] at 70°C. Activity unit is defined as 1 μmol maltose liberated per minute at the optimum temperature.

Gradient SDS polyacrylamide gels were prepared and run according to Laemmli and Favre [17] and silver stained [18].

RESULTS

Cloning and Sequencing of the Gene Coding for α-Amylase

A λ EMBL3 [8] gene library was constructed in E. coli from chromosomal DNA of B. stearothermophilus partially cleaved with Sau3AI and size fractionated. Recombinant phages were plated with soft agar containing 1% corn starch in order to screen α-amylase. Among 7300 plaques screened, eight produced a detectable halo (i.e., a zone of clearance) after staining the plates with iodine. DNA from one of the α-amylase hybrid phages was cleaved either partially with Sau3AI or totally with PvuII and HindIII and ligated with BamHI cleaved pUC-8 or PvuII and HindIII cleaved pBR322, respectively. After transformation, E. coli JM103 cells were plated on L-plates supplemented with ampicillin (100 μg/ml) and 1% corn starch.

For sequencing we initially selected the pUC-8 based subclone, designated pCSS1 (see Figure 1A), which produced most α-amylase as detected on starch plates. Sequence analysis (see below) showed that a rearrangement had occurred in the subcloning step in 5'-flanking region of the gene in pCSS1. In the pBR322 based construction, called pCSS4 (see Figure 1B), the PvuII-HindIII insert is continuous, which was verified by restriction mapping and partial sequence analysis. Sequencing (see Figure 1A for strategy) of the α-amylase gene was performed using the chain termination method [14]. The nucleotide sequence and the deduced amino acid sequence are presented in Figure 2.

5'Flanking sequence reveals two possible promoter regions (in addition to the nearby lac-promoter) in positions -145 to -115 and -50 to -25 from the GTG initiation codon. The more distant region (P_1) consists of a consensus TATAAT -10 region and a -35 region ATGGCA (two mismatches from consensus TTGACA) separated by 19 nucleotides. The region proximal to the initiation codon (P_2) consists of a -10 region TAACAT (2 mismatches) and a

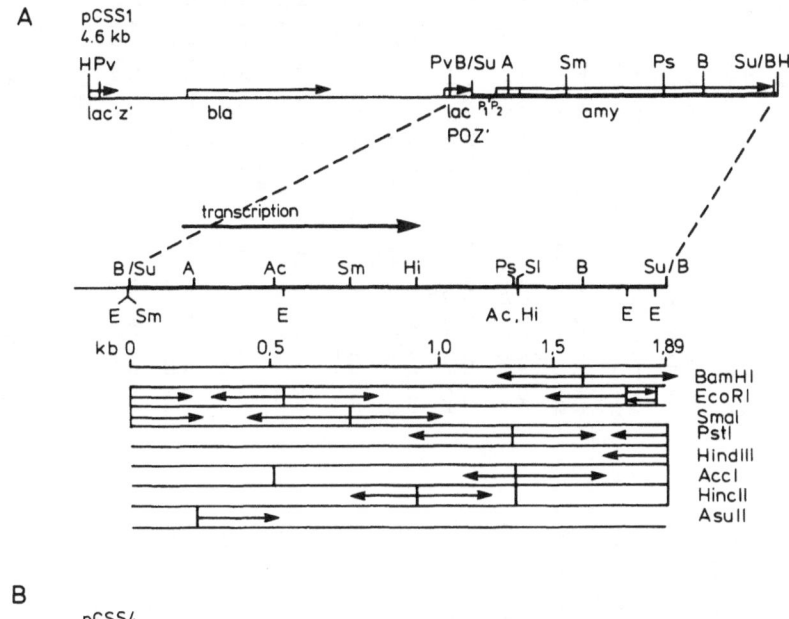

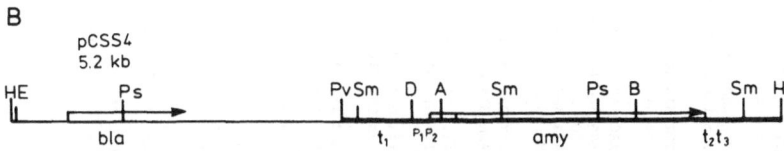

Fig. 1.   Restriction maps of pCSS1 and pCSS4 and sequencing strategy for B. stearothermophilus α-amylase gene.   (A) pCSS1 is a pUC-8 derivative where the α-amylase gene has been cloned from a λ EMBL3 hybrid by partial Sau3AI digestion. The thick line is the insert, which is enlarged below showing the more detailed restriction map. Sequencing strategy is presented below it with the arrows.   (B) pCSS4 is a pBR322 derivative. pBR322 was cleaved with PvuII and HindIII and the 2.3 kb fragment was isolated (the thin line).   An α-amylase carrying hybrid phage λ EMBL3 was cleaved with PvuII and HindIII, a 3 kb fragment containing the α-amylase gene was isolated (the thick line) and the two fragments were ligated together.
Abbreviations of the enzymes: A - AsuII, AC - AccI, B - BamHI, D - DraI, E - EcoRI, H - HindIII, Hi - HincII, Ps - PstI, Pv - PvuII, Sl - SalI, Sm - SmaI, Su - Sau3AI.
Other symbols: lacPOZ' - E. coli lac promoter, operator and a 5'-fragment of β-galactosidase gene, bla - β-lactamase gene, amy - α-amylase structural gene, t_1 - transcription terminator preceding the α-amylase gene, t_1, t_2- transcription terminator for the α-amylase gene.

Fig. 2 — Nucleotide sequence and deduced amino acid sequence of *B. stearothermophilus* α-amylase gene.

```
GATC ATATTTTGCA TCGGTCCTGG CGTCATCCA CCCTACTCCC CGATCACCCG CACGGTGAAA GTCACC -165

AATG TGGGACGTCC GTCGATGGCA GAAGATCACA AATAAAAATT ATAATAGACG TAACCGTTCG AGGTTT  -95
          P1(-35)                                P1(-10)

TGCT TCCTGTTTAC TCTTTTTATG CAATCATTTC CCTTCATTTT TTGGAATCCA AACCGTCGAA TGTAAC  -25
               P2(-35)                       P2(-10)            P2

ATTT GATTAGGGGG GAAGGGCATT GTG CTA ACG TTT CAC ATC ATT CGA AAA GGA TGG            36
     (-10)      S-D         fMet Leu Thr Phe His Ile Ile Arg Lys Gly Trp
                            -34
                                      -20

ATG TTC CTC GCG TTT CTC ACT GCC TTG CTC CTG CTG TGC CCA ACC GGA CAG CCC          93
Met Phe Leu Ala Phe Leu Leu Leu Leu Cys Pro Thr Gly Gln Pro
                                           -1 +1

GCC AAG GCT GCC GCA CCG TTT AAC GGC ACC ATG GCG ATG TTT GAA TGG TAC TTG        150
Ala Lys Ala Ala Ala Pro Phe Asn Gly Thr Met Ala Met Phe Glu Trp Tyr Leu
         ↑         10                          20                          30

CCG GAT GAT GGC ACG TTA TGG ACC AAA GTG GCC AAT GAA GCA AAC AAT TTA TCC AGC AGC  207
Pro Asp Asp Gly Thr Leu Trp Thr Lys Val Ala Asn Glu Ala Asn Asn Leu Ser Ser
                       40                                  50

CTT GGC ATC GCT CTT TGG CTG CCG CCC GCT TAT AAA GGA ACA AGC AGC GAC GAC        264
Leu Gly Ile Ala Leu Trp Leu Pro Pro Ala Tyr Lys Gly Thr Ser Arg Ser Asp
                  60                                70

GTA GGG TAC GGA GTA TAC GAC TTG TAT GAT CTC GGT GAA TTC AAT CAA AAA GGG GCC    321
Val Gly Tyr Gly Val Tyr Asp Leu Tyr Asp Leu Gly Glu Phe Asn Gln Lys Gly Ala
         80                                90

GTC CGC ACA AAA TAC GGA ATG ATC CAA GCT CAA GTG TTC CAA ATT CAA AAA GCC GCC GAC  378
Val Arg Thr Lys Tyr Gly Met Ile Gln Ala Gln Val Phe Gln Leu Gln Ala Ala His
         100                                110

GCC GCT GGA ACA TAC AGC GAC GTG GTG TAC GAC CAT TTC GAC TGG GGC GGC GAC      435
Ala Ala Gly Thr Tyr Ser Asp Val Val Tyr Asp His Phe Asp Trp Gly Gly Asp
         120                                130

GGC ACG GAA TGG GTG GTG CGC GTC AAT CCG TCC AAC CGC AAC CAA GAA ATC          492
Gly Thr Glu Trp Val Val Arg Val Asn Pro Ser Asn Arg Asn Gln Glu Ile
                  130                                    140

TCG GGC ATC TAT CAA ATC CAA GCA TGG ACG AAA TTT GAT TTT CCC GGG CGG GGC AAC    549
Ser Gly Ile Tyr Gln Ile Gln Ala Trp Thr Lys Phe Asp Phe Pro Gly Arg Gly Asn
     150                                160

ACC TAC TCC AGC TTT AAG TGG CGC TGG TAC CAT TTT GAT GGC GTT GAT TGG GAC GAA    606
Thr Tyr Ser Ser Phe Lys Trp Arg Trp Tyr His Phe Asp Gly Val Asp Trp Asp Glu
     170                                180

AGC CGA AAA TTG AGC CGC ATT TAC AAA TTC CGC GGC ATC GGC AAA GCG TGG GAT TGG    663
Ser Arg Lys Leu Ser Arg Ile Tyr Lys Phe Arg Gly Ile Gly Lys Ala Trp Asp Trp
     190                                200

GAA GTA GAC ACG GAA AAC GGA AAC TAT GAC TAC TTA ATG TAT GCT GAT CTT GAT ATG    720
Glu Val Asp Thr Glu Asn Gly Asn Tyr Asp Tyr Leu Met Tyr Ala Asp Leu Asp Met
     210                                220

GAT CAT CCC GAC GTC GTG ACT GAC GTG CTA AAA GCT GGG AAA GTC TAT GAT TGG GTC AAC ACA  777
Asp His Pro Asp Val Val Thr Asp Val Leu Lys Ala Gly Lys Val Tyr Asp Trp Val Asn Thr
     220
```

```
AGA ACA CTT GAT GGG TTC CGG CTT GAT GCC GTC AAG CAT ATT AAG TTC AGT TTT TTT    834
Arg Thr Leu Asp Gly Phe Arg Leu Asp Ala Val Lys His Ile Lys Phe Ser Phe Phe
                                             240                    260

CCT GAT TGG TTG TCG TAT GTG CGT TCT CAG CGC AAG CGC CTA TTT ACC GTT GGG        891
Pro Asp Trp Leu Ser Tyr Val Arg Ser Gln Arg Lys Arg Leu Phe Thr Val Gly
                   250                              270            280

GAA TAT TGG AGC TAT GAC ATC AAC AAG TTG CAC AAT TAC ATT ATG AAA ACA GGA        948
Glu Tyr Trp Ser Tyr Asp Ile Asn Lys Leu His Asn Tyr Ile Met Lys Thr Gly
                                   290                          300

ACG ATG TCT TTG TTT GAT GCC CCG TTA CAC AAC AAA TTT TAT TAT GCT TCC AAA TCA   1005
Thr Met Ser Leu Phe Asp Ala Pro Leu His Asn Lys Phe Tyr Tyr Ala Ser Lys Ser
                                        310                          320

GGG GGC ACA TTT GAT ATG CGC ACG TTA ATG ACG AAT ACT CTC ATG AAA GAT CAA CCA   1062
Gly Gly Thr Phe Asp Met Arg Thr Leu Met Thr Asn Thr Leu Met Lys Asp Gln Pro
                          330

ACA TTG GCC GTC GTT GAT AAT CAT GAC ACC CAA CGG CTG CAG                        1119
Thr Leu Ala Val Thr Phe Val Asp Asn His Asp Thr Gln Arg Leu Gln
340                          350

TCA TGC GTC GAC TGG TTC AAA CCG TTC GCT TAT GCC TTT ATT CTA ACT CGG CAG       1176
Ser Trp Val Asp Trp Phe Lys Pro Phe Ala Tyr Ala Phe Ile Leu Thr Arg Gln
360                          370

GAA GGA TAC CCG TGC GTC TTT TAT GGT GGA TTC TAC GGC ATT CCA CAA TAT GCT AAC ATT  1233
Glu Gly Tyr Pro Cys Val Phe Tyr Gly Asp Tyr Tyr Gly Ile Pro Gln Tyr Asn Ile
                       380               390

CCT TCG CTG AAA AGC ATC ATC GAT CCG CTC CTC GAC AGG AGG TAT GCT TAC GGG       1290
Pro Ser Leu Lys Ser Ile Ile Asp Pro Leu Leu Asp Arg Arg Tyr Ala Tyr Gly
400                          410

GGA CAA CAT GAT TAT CTT GAC CAC TCC GAC ATC ATC GGG TGG ACA AGG GAA GGG GGA   1347
Gly Thr Gln His Asp Tyr Leu Asp His Ser Asp Ile Ile Gly Trp Thr Arg Glu Gly Gly
420                          430

GTC ACT GAA AAA CCA GGA TCC GGA CTG CTG GCA CAC AAC GTG TTC TAT GAC CTT ACC   1404
Val Thr Glu Lys Pro Gly Ser Gly Leu Ala Ala His Gln Lys Val Phe Tyr Asp Leu Thr
440                          450

AGC AAA TGG ATG TAC GTT CCC GGG CGG GGA AAA GTG TTC TAT GAC CTT ACC          1461
Ser Lys Trp Met Tyr Val Gly Gln His Ala Gly Lys Thr Asn Ser Asp Trp Ser Pro Gly Gly
                                             460                          470

GGC AAC CGG AGT GAC ACC GTC AAC ATC AAC AGT GAT GGA TGG GGG GAA TTC AAA GTC   1518
Gly Asn Arg Ser Asp Thr Val Asn Ile Asn Ser Asp Gly Trp Gly Glu Phe Lys Val
480

AAT GGC GGT TCG GTT TCC GTT TGG GTG CCT AGA ACG ACG ACG CTG TCT ATC GCT GCT   1575
Asn Gly Gly Ser Val Ser Val Trp Val Pro Arg Lys Thr Thr Val Ser Thr Ile Ala
490                          510

TGG TCG ATC ACA ACC CGA TGG CAG ACT GAT GAA TTC GTC CGT TGG ACC CGA CCA CGG   1632
Trp Ser Ile Thr Thr Arg Pro Trp Thr Asp Glu Phe Val Arg Trp Thr Glu Pro Arg
                  500                          515

TTG GTG GCA TGG CCT TGA TGCCTGC                                                1657
Leu Val Ala Trp Pro END
```

Fig. 2. Nucleotide sequence and deduced amino acid sequence of B. stearothermophilus α-amylase gene. The nucleotide sequence of the insert in pCSS1 is shown. Numbering of the nucleotides, indicated at the end of each line, begins at the first base of the translation initiation codon GTG. Numbering of the amino acids, indicated above the corresponding codons, begins at the first amino acid (Ala1) of mature α-amylase. Possible promoter regions (P1 and P2) and ribosome binding site (S-D) are underlined. The thick arrow indicates the correct signal peptidase cleavage site in E. coli (see text).

-35 region TTGGAA (2 mismatches) separated by 16 nucleotides. P_1 is more likely to act as a promoter, at least in E. coli, since deletion of the 5' upstream region, beginning at Sau3AI site at position -137 between -35 and -10 regions of P_1, almost totally abolishes α-amylase production in a pBR322 based construction (not shown). Preceding the promoter region there is an inverted repeat, which can form a fairly stable stem and loop structure (t_1 in pCSS4 map, not shown in the sequence). This is probably a transcription terminator of an upstream gene.

Preceding the initiation codon there is a region (AGGGGGAAG) which is partially complementary to the 3'-end of B. stearothermophilus 16S rRNA [19]. This region closely resembles those described for Bacillus ribosome binding sites [20]. Although there are many ATG codons in the 5'flanking region of the gene, the GTG beginning at +1 (Figure 2) is the only one that precedes a typical signal peptide like sequence [21,22]. For this reason we propose that GTG is used as the translational initiation codon. After the termination codon two inverted repeats are present, which probably act as the transcription terminator for α-amylase gene.

The amino acid sequence of mature α-amylase of B. stearothermophilus is fairly homologous with two other liquefying α-amylases from B. licheniformis [23] and B. amyloliquefaciens [24].

Processing of B. Stearothermophilus α-Amylase in B. Stearothermophilus and in E. Coli

The open reading frame is 1650 bp, which codes for a preprotein of MW 63,000. The signal peptidase cleavage site was determined by manual Edman degradation (4 cycles) [24,25] for purified α-amylases produced by both the donor strain B. stearothermophilus ATCC 12980 and E. coli JM103 (pCSS1). The cleavage site for B. stearothermophilus lies between amino acids 34 (Ala) and 35 (Ala). E. coli cleaves 60% of the molecules at the same site and 40% between residues 31 (Pro) and 32 (Ala). The signal peptide is thus 34 amino acid residues long. The results from SDS polyacrylamide gels for the enzyme from both B. stearothermophilus and E. coli (pCSS1) (Figure 3) are identical with the calculated molecular weight for the mature α-amylase.

Secretion of α-Amylase into the Growth Medium by E. Coli

When E. coli JM103 carrying either pCSS1 or pCSS4 were grown on starch plates a clear difference in the sizes of haloes was detected. After overnight growth JM103 (pCSS1) produced much larger haloes than JM103 (pCSS4). To find out if this is due to differences in gene dosage (i.e., plasmid copy numbers) and to the fact that pCSS1 has a tandem promoter (lacP and amyP) in front of the gene, we grew both clones at 37°C in LB medium supplemented with ampicillin and followed the kinetics of α-amylase production by measuring enzyme activities from culture supernatants and from cells (Figure 4). (Fractionation of cells by osmotic shock treatment revealed that the total cell bound enzyme activity is about the same as that found in the periplasm.) β-Lactamase was chosen as a marker for periplasmic enzymes. E. coli harboring only the vector plasmid pUC-8 retains β-lactamase activity entirely in the periplasm (Figure 4E), where it is degraded relatively soon after the onset of stationary phase. In JM103 (pCSS1) β-lactamase is initially periplasmic (Figure 4A), but is also present in the growth medium from the beginning of stationary phase. β-Lactamase found in the medium is not degraded anymore. In JM103 (pCSS1) α-amylase (Figure 4B) is constantly present in the periplasm, but ca. 25% of the total activity is in the medium after 24 hour growth. (The extracellular activity visible at the beginning is due to residual activity from the inoculum of an overnight culture.) In JM103 (pCSS4) β-lactamase pro-

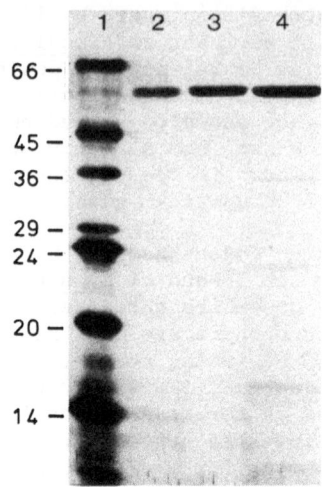

Fig. 3.   SDS polyacrylamide gel electrophoresis of purified α-amylases.
α-Amylase was purified from stationary phase culture supernatants
of E. coli JM103(pCSS1) (lane 2.), B. subtilis 1A150(pCSS502)
(carries the B. stearothermophilus α-amylase, unpublished) (lane
3.) and B. stearothermophilus ATCC12980 (lane 4.).   A 5 - 20%
gradient gel was prepared according to Laemmli and Favre [17], and
silver stained [18].   Molecular weight markers are (lane 1.)
Bovine Albumin, 66 kD, Egg Albumin, 45 kD, Glyceraldehyde-3-
Phosphate Dehydrogenase, 36 kD, Carbonic Anhydrase, 29 kD,
Trypsinogen, 24 kD, Trypsin Inhibitor, 20 kD and α-Lactalbumin,
14 kD.

duction (Figure 4C) is much smaller due to the lower copy number of the
vector and the activity remains totally periplasmic.   α-Amylase produced by
this strain (Figure 4D) remains also periplasmic.   The final total amount
of α-amylase is also lower than in JM103 (pCSS1).   The secretion of
α-amylase and other periplasmic enzymes from the E coli periplasm is even
more pronounced in JM103 (pCSS1) cultures when the lac promoter has been
induced by IPTG, causing further increase in α-amylase production (not
shown).

DISCUSSION

    The previously reported sequences of B. stearothermophilus α-amylase
gene show some differences with each other as well as with the sequence
published here: notably at positions (Figure 1B) 779, 784, 749 and 750
compared to Nakajima et al. [6] and at positions 68, 89, 90, 92, 227, 319,
432, 779-785, 806 and 808 compared to Ihara et al. [5].   The major difference is, however, at the 5' upstream region (the Sau3A fragment -234 -
-137) including the -35 region of the P_1 promoter in pCSS1.   This discrepancy originates from a rearrangement in the subcloning step, since we
chose for sequencing the recombinant which produced most amylase (as
detected on starch plates by iodine staining).   However, this does not
affect the strength of the promoter, since at least as much α-amylase is
produced by JM103 (pCSS1) as by similar constructions, where the promoter
area is intact (not shown).   In pCSS4 the 5'-flanking region is intact
(Figure 1B), but the copy number is lower due to the vector used.   The
other possible promoter region (P_2 in Figure 2) seems not to be active (or
at least very weak) in El coli, since deletion of the -35 region from P_1
abolishes α-amylase production almost totally (not shown).

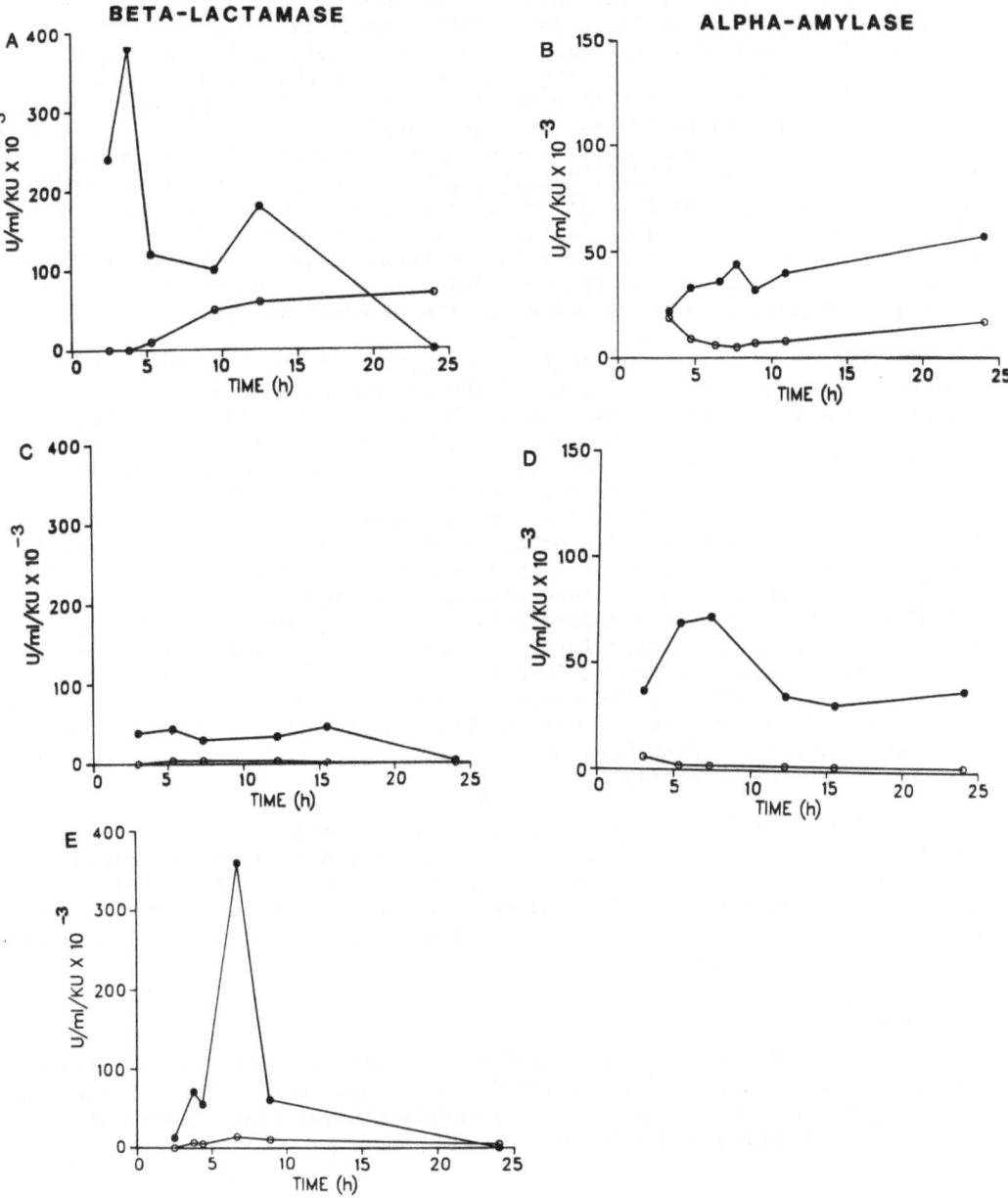

Fig. 4.   Secretion of α-amylase and β-lactamase by E. coli. E. coli JM103
carrying either pCSS1 (A and B), pCSS4 (C and D) or pUC-8 (E) were
grown in LB medium supplemented with ampicillin (100 ug/ml) at
37°C.   Samples (1 ml) were taken as indicated, the cells were
separated from the medium by centrifugation, suspended in 1 ml of
fresh cold medium and broken by sonication. β-lactamase (A, C and
E) and α-amylase (B and D) activities were measured from aliquots
of the broken cells (solid circles) and the culture supernatants
(open circles).   Enzyme activities are indicated as units per 1 ml
of culture per Klett$_{62}$ unit (KU).

The signal peptide, as determined by amino terminal amino acid sequence analysis for secreted enzymes, was shown to be 34 residues long in B. stearothermophilus and 34 residues (60%) and 31 residues (40%) long in E. coli. Its amino terminal region (N region as defined by von Heijne [26]) is 11 residues long carrying four positively charged amino acids. This is followed by a hydrophobic region (H region) of 14 residues. The C or cleavage region (9 residues) is exceptionally long [26]. The reason for ambiguous processing by E. coli is not clear at present, but since the C-region is longer than usual, E coli might try to "shorten" it by finding the next suitable processing site which matches the consensus length of the C-region, which is 6 residues instead of 9. On the other hand, the new cleavage site is not a typical one for bacteria [21,26], but some similar sites have been found in eukaryotes. This, as well as other features of the signal sequence, is now studied by site directed mutagenesis.

E. coli is not usually thought to secrete proteins into the growth medium, except for some exceptional proteins such as enterotoxins, colicins and hemolysin of certain strains [27]. However, secretion or leaking of cloned enzymes from E. coli has been observed also in other laboratories recently [28,29]. α-Amylase activity found in the stationary phase culture medium (Figure 4A) is not a consequence of cell lysis, since no decrease in OD was observed even after 24 hour growth. Also, when transformed to E coli HB101, where kinetics of production of α-amylase and β-lactamase are similar to JM103, no leakage of β-galactosidase was observed (not shown). Based on these results we conclude that accumulation of α-amylase in the periplasm of E. coli causes alterations in the outer membrane allowing periplasmic enzymes to escape into the growth medium. These alterations have not yet been characterized further, but since leakage occurs only after sufficient amount of the enzyme is produced, one possible explanation is that α-amylase binds to an outer membrane component titrating it out from the membrane or disturbing membrane structure causing the membrane to lose its permeability barrier.

These observations may prove useful for production of other foreign proteins in E. coli. Other genes can be either fused to the α-amylase gene or cloned to the same vector and when the new protein is translocated into the periplasm, simultaneous (or induced) accumulation of α-amylase releases the contents of the periplasm into the growth medium. This possibility is now under investigation.

Acknowledgements

We wish to thank Peter Meyer and Matti Lähde for technical assistance and Nisse Kalkkinen and Marc Baumann for amino acid sequencing. Financial support from Neste Oy foundation and Technology Development Center of Finland is gratefully acknowledged.

REFERENCES

1. F. G. Priest, Bacteriol. Rev., 41:711 (1977).
2. S. Aiba, K. Kitai, and T. Imanaka, Appl. Environ. Microbiol., 46:1059 (1983).
3. J. R. Mielenz, Proc. Nat. Acad. Sci., USA, 80:5975 (1983).
4. N. Tsukagoshi, H. Ihara, H. Yamagata, and S. Udaka, Mol. Gen. Genet., 193:58 (1984).
5. H. Ihara, T. Sasaki, A. Tsuboi, H. Yamagata, N. Tsukagoshi, and S. Udaka, J. Biochem., Tokyo, 98:95 (1985).
6. R. Nakajima, T. Imanaka, and S. Aiba, J. Bacteriol., 163:401-406 (1985).

7. W. Arber, L. Enquist, B. Hohn, N. E. Murray, and K. Murray, in: "Lambda II", R. W. Hendrix, J. W. Roberts, F. W. Stahl and R. A. Weisberg, eds., Cold Spring Harbor Laboratory, Cold Spring Harbor, New York, p. 433 (1983).

8. A.-M. Frischauf, H. Lerach, A. Poustka, and N. Murray, J. Mol. Biol., 170:827 (1984).

9. J. Vieira, and J. Messing, Gene, 19:259 (1982).

10. J. Messing, Methods Enzymol., 101:20 (1983).

11. T. Maniatis, E. F. Fritsch, and J. Sambrook, "Molecular Cloning", Cold Spring Harbor Laboratory, Cold Spring Harbor, New York (1982).

12. P. S. Lovett, and K. M. Keggins, Methods Enzymol., 68:342 (1979).

13. A. I. Suominen, M. I. Karp, and P. I. Mäntsälä, Biochem. Int., 8:209 (1984).

14. F. Sanger, S. Nicklen, and A. R. Coulson, Proc. Nat. Acad. Sci., USA, 74:5463 (1977).

15. M. O. Biggin, J. J. Gibson, and G. J. Hong, Proc. Nat. Acad. Sci., USA, 80:3963 (1983).

16. B. Bernfeld, Methods Enzymol., 1:149 (1955).

17. U. K. Laemmli, and M. Favre, J. Mol. Biol.., 80:575 (1973).

18. B. L. Nielsen, and L. R. Brown, Anal. Biochem., 141:311 (1984).

19. J. Shine, and L. Dalgarno, Nature, 254:34 (1975).

20. J. R. McLaughling, C. L. Murray, and J. R. Rabinowitz, J. Biol. Chem., 256:11283 (1981).

21. A. I. Suominen, and P. Mäntsälä, Int. J. Biochem., 15:591 (1982).

22. M. E. E. Watson, Nucleic Acids Res., 12:5145 (1984).

23. T. Yuuki, T. Nomura, H. Tezuka, A. Tsuboi, H. Yamagata, N. Tsukagoshi, and S. Udaka, J. Biochem.., Tokyo, 98:1147 (1985).

24. K. Takkinen, R. F. Pettersson, N. Kalkkinen, I. Palva, H. Söderlund, and L. Kääriäinen, J. Biol. Chem., 258:1007 (1983).

25. J. D. Peterson, S. Nehrlich, P. E. Oyer, and D. F. Steiner, J. Biol. Chem., 247:4866 (1972).

26. G. von Heijne, J. Mol. Biol., 184:99 (1985).

27. A. P. Pugsley, and M. Schwartz, FEMS Microbiol. Rev., 32:3 (1985).

28. N. R. Gilkes, D. G. Kilburn, R. C. Miller Jr., and R. A. J. Warren, Biotechnol., 2:259 (1984).

29. T. Kobayashi, C. Kato, T. Kudo, and K. Horikoshi, J. Bacteriol., 166:728 (1986).

EXOCELLULAR PROTEIN AND α-AMYLASE SECRETION

IN <u>BACILLUS SUBTILIS</u>

J. Pazlarová

Department of Enzyme Engineering, Institute of
Microbiology, Czechoslovak Academy of Sciences
Prague 4, Czechoslovakia

The production and secretion of α-amylase (EC 3.2.1.1) was studied as a model for protein secretion in <u>Bacillus subtilis</u> during the exponential phase of growth. The synthesis of α-amylase in this strain is dependent on the presence of plasmid pMI10 in the cell [1]. The enzyme itself belongs to the group of well characterized extracellular bacillar enzymes [2]. It is well known that the synthesis and secretion of these types of enzymes takes place mainly after the end of exponential growth [3,4]. Evaluation of the glucose addition (catabolite repression) and the effect of the growth rate on α-amylase secretion was the aim of this study.

We followed the progress of secretion of total extracellular proteins by <u>B. subtilis</u>, grown aerobically at 30°C in a simple mineral-starch medium [5] supplemented with casamino acids. ^{14}C-leucine in concentration 0.1 μCi/ml served as a marker for both intracellular (whole cells) and extracellular protein synthesis. During the followed period (210-300 min) the proportion between the intracellular and the extracellular protein was linear. The addition of glucose at different times had no effect on the character of the secretion (Figure 1). The specific growth rate (μ) under these conditions remained constant ($\mu = 0.55$ h^{-1}). When the amount of casamino acids in the medium was decreased from 2 mg/ml is 0.2 mg/ml a biphasic pattern of the extracellular protein secretion appeared (Figure 2). Under these conditions (0.2 mg casamino acids/ml) the exoprotein secretion in the end of the 210 min period was equivalent to 9,5% of the total bacterial protein. The control cells (2 mg casamino acids/ml) produced 4% of the total bacterial protein as an exoprotein. The elevation of the exoprotein secretion in the second phase was connected with the decrease of the specific growth rate below $\mu = 0.40$ h^{-1}. In previous experiments performed in a chemostat [6], the optimal μ for α-amylase synthesis was between 0.2-0.3, the μ_{max} being higher than 0.45. A complex medium was used in that case.

Electrophoretic analysis of the exoprotein at the end of the experimental period (210 min) indicated the presence of one protein only. The culture supernatant after 210 min growth of the cells was freeze-dried, the solids were diluted in a 100 time smaller amount of distilled water. Figure 3 presents the analysis of this solution in 7,5% polyacrylamide gel in presence of SDS [7].

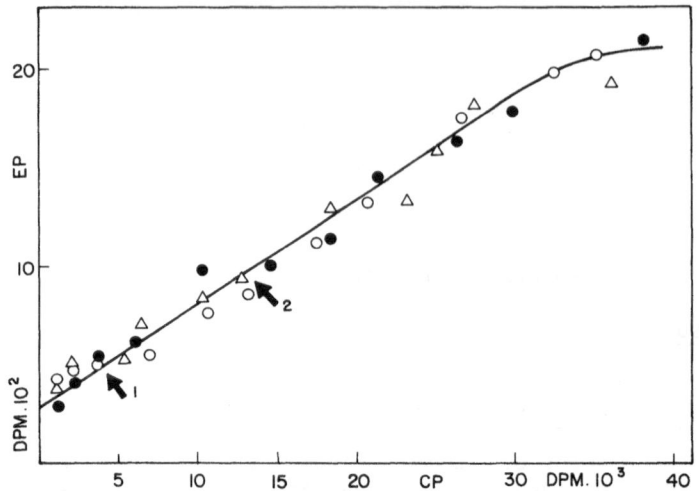

Fig. 1. Protein secretion relative to total protein synthesis in Bacillus subtilis in a medium supplemented by glucose. The culture was grown aerobically at 30°C. An overnight culture was diluted to an optical density of 0.2 (650 nm) in a medium containing 2 mg casamino acids/ml. After 20 min of growth ^{14}C-leucine was added. The duration of the experiment was 300 min. Arrows indicate the addition of glucose to a final concentration of 5 mg/ml. CP, cell protein: 0.5 ml of culture broth was precipitated with 0.5 ml 10% TCA, the precipitate was collected on Synpor No. 6 membrane filter (pore size 0.4 μm Synthesia, Czechoslovakia) and counted on Beckman LS 9000. EP, extracellular protein: 5 ml of culture broth was centrifuged. To 4 ml of the supernatant was added 0.1 ml 1% casein solution and 0.5 ml 50% TCA. The precipitate was collected on Synpor No. 6 and counted. Sample interval was 15 min. o, control cells without glucose addition, ●, addition of glucose at 60 min, △, addition of glucose at 150 min.

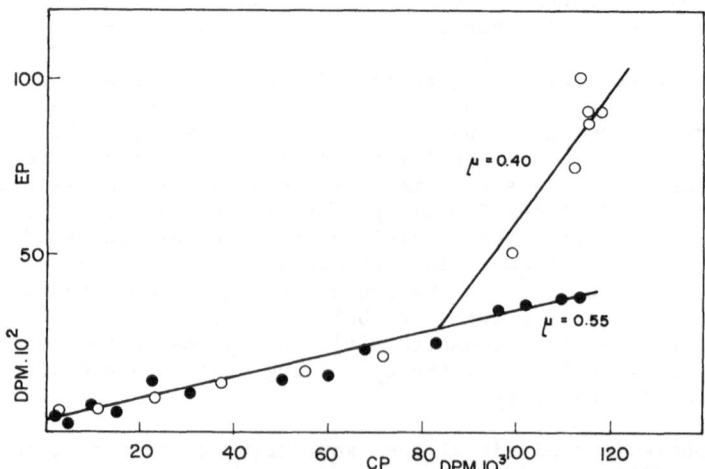

Fig. 2. Secretion of Bacillus subtilis protein into the medium with different concentration of casamino acids. Cultivation conditions and samples were identical as described in Figure 1. The duration of the experiment was 210 min. Growth rate was monitored from absorbance values obtained at 650 nm, taken at 30 min intervals. o, 0.2 mg casamino acids/ml, ●, 2 mg casamino acids/ml.

140

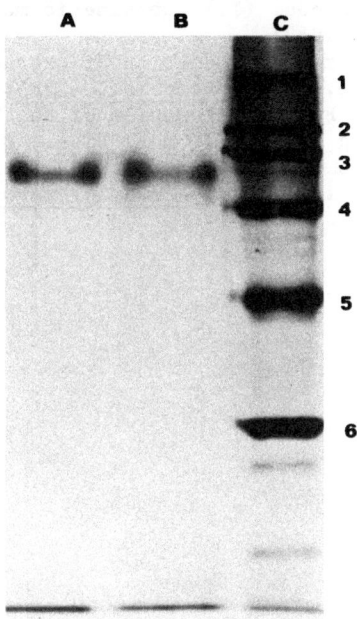

Fig. 3. SDS-polyacrylamide gel electrophoresis of extracellular protein.
Detection was with Coomassie Blue. Lanes A and B, sample containing 2.95 mg protein/ml and 70 U/ml of α-amylase estimated by
3,5-dinitrosalicylic acid [8]. Lane C, 1 myosin (205,000), 2
β-galactosidase (116,000), 3 phosphorylase (97,000), 4 bovine
serum albumine (66,000), 5 egg serum albumine (45,000), 6 carbonic
anhydrase (29,000). The author thanks Dr. M. Jiresová for per-
forming the electrophoretic analysis of α-amylase.

CONCLUSION

Under the conditions used in the experiments, the concentration of
casamino acids was the growth limiting factor. When the concentration was
2 mg casamino acids/ml the μ was equal to 0.55 h^{-1} and protein secretion
was proportional to the overall protein synthesis rate. After exhaustion
of amino acids the growth rate decreased and protein secretion increased
2-fold.

REFERENCES

1. P. Tichý, J. Pazlarová, M. Hartmann, Z. Fencl, L. Erbenová, O. Benada,
    and V. Krumphanzl, Isolation of the pMI 10 Plasmid from the α-
    amylase producing strain of Bacillus subtilis, Mol. Gen. Genet.,
    181:248 (1981).
2. F. G. Priest, Extracellular enzyme synthesis in the genus Bacillus,
    Bacteriol. Rev., 41:711 (1977).
3. G. Coleman, The effect of glucose on the differential rates of extra-
    cellular protein and α-toxin formation by Staphylococcus aureus
    (Wood 46), Arch. Microbiol., 134:208 (1983).
4. S. Kinoshita, Bacterial α-amylase, in "Annual Reports of International
    Center of Cooperative Research and Development in Microbial Engin-
    eering", Faculty of Engineering, Osaka University, Osaka, Japan
    (1980).
5. G. Terui, M. Okazaki, and S. Kinoshita, Kinetic studies on enzyme

production by microbes. (I.)   On kinetic models, J. Ferment. Technol., 45:497 (1967).

6. A. Fencl and J. Pazlarová, Production of extracellular enzymes in continuous culture, Folia Microbiol., 27:340 (1982).

7. U. K. Laemmli and M. Favre, Maturation of the head of bacteriophage T4. I. DNA packing events, J. Mol. Biol., 80:575 (1973)

8. P Bernfeld, Amylases α and β, in "Methods in Enzymology" S.P. Colowick, N.O. Kaplan, eds., Academic Press, New York, pp.149, (1955).

EXTRACELLULAR AMYLASE FROM THE THERMOPHILIC

FUNGUS <u>THERMOMYCES LANUGINOSUS</u>

B. Jensen, J. Olsen and K. Allermann

Department of General Microbiology
University of Copenhagen
Sølvgade 83H, DK-1307, Copenhagen K, Denmark

INTRODUCTION

The thermophilic fungus <u>Thermomyces lanuginosus</u> formerly named <u>Humicola lanuginosa</u> (Domsch et al., 1980) appears to be an excellent test organism for the study of extracellular amylase from a thermophilic fungus (Barnett and Fergus, 1971). Basaveswara Rao et al. (1979) have identified a glucoamylase in the growth medium containing starch from static cultures as well as shake flask cultures. The yield of activity was found to be 2.5 fold greater in shake cultures than in static cultures. Adams (1981) related amylase production to mycelial growth in static cultures and proposed that a mycelial bound amylase could account for the rapid disappearance of starch from the medium and that the extracellular amylase might in large part be due to autolysis. In this paper the relation between growth, enzyme production and substrate utilization was studied in a stirred laboratory fermenter with three different carbon sources.

MATERIALS AND METHODS

<u>Microorganism</u>: <u>Thermomyces lanuginosus</u> (single spore isolate from strain No. 1457. The Royal Veterinary and Agricultural University, Copenhagen) was maintained on agar plates at 50°C between recultivation.

<u>Media</u>: All conc. in g/l. Yeast extract 4.0; $K_2HPO_4.3 H_2O$ 1.0; $MgSO_4.7 H_2O$ 0.5 was dissolved in 3/4 distilled and 1/4 tap water; pH was adjusted to 6.5. As carbon source either starch, glucose or maltose at a concentration of 15 g/l was added to this medium.

<u>Inoculum</u>: Spores from two agar plates were suspended in 100 ml 0.05 (v/v) Triton X-100 and filtered through four layers of sterile gauze, before they were pumped into the fermenter.

<u>Culture conditions</u>: Batch fermentations were performed in a 2 1 fermenter, type LH-500, LH Fermentation Ltd., UK, with mechanically driven impellers. The fungus was cultivated at 50°C with a working volume of 1.8 1, an air flow rate of 1 VVM, and a rotational speed of the impeller of 600 to 1000 rpm. The dissolved oxygen concentration of the medium, which was followed during growth, did not fall below 20% saturation.

Analytical procedures: Protein was determined with the Biorad-method. Glucose was determined by the hexokinase-method (Boehringer), amylase activity by the method of Manning and Campbell (1961) at 60°C. One unit of amylase is defined as the amount of protein which will hydrolyse 10 mg of starch per minute.

RESULTS AND DISCUSSION

Thermomyces lanuginosus was able to grow and produce amylases with all 3 carbon sources (Figures 1, 2 and 3).

When the fungus was grown with 15 g/l of starch (Figure 1), this polymer was broken down (data not shown) within the first day after inoculation, and glucose was found in the medium until the stationary phase was reached on day two. With maltose as the carbon source glucose appeared in the medium on day two (Figure 2).

Protein and amylase activity were excreted during growth (Figures 1, 2 and 3), but the amylase activity mainly late in the growth phase, when the glucose concentration was low.

The data shown for dry weight was only from suspended mycelium and must not be regarded as data for the total biomass in the fermenter, as heavy wall growth always took place. The total amount of biomass (suspended and wall-grown) at the end of the fermentations was similar with all three media (2.8 - 3.6 g/l).

The data from these experiments did not give a clear-cut picture as to the origin of the extracellular amylase activity. Some autolysis did take place during the stationary phase, but the excretion of protein was clearly growth related. The liberation of glucose (Figures 1 and 2) indicated that glucoamylase was part of the amylase activity. Preliminary experiments (unpublished) with gel electrophoresis have shown more than one protein with amylase activity.

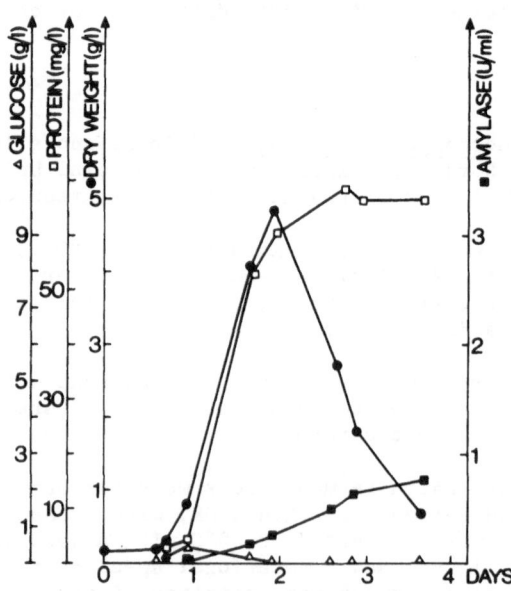

Fig. 1. Growth (●) and changes in extracellular glucose (△), protein (□) and amylase (■) during fermentation with starch as carbon source.

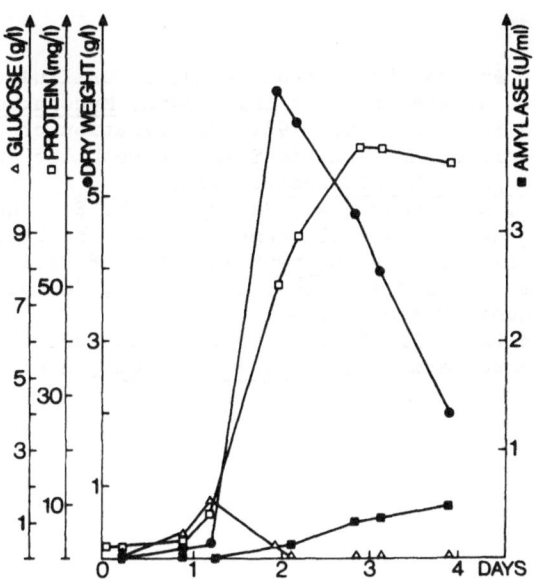

Fig. 2. Growth (●) and changes in extracellular glucose (△), protein (□) and amylase (■) during fermentation with maltose as carbon source.

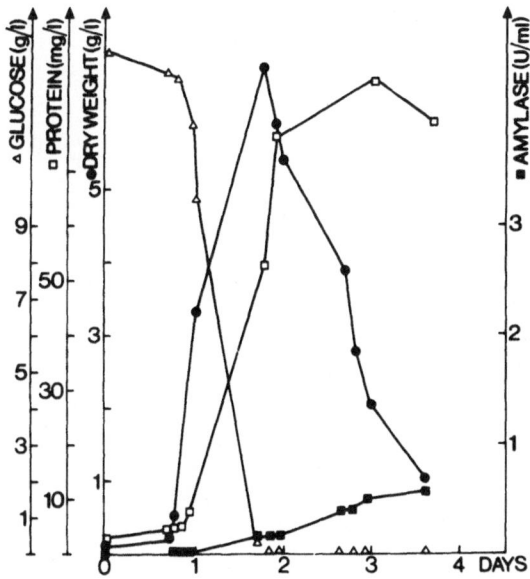

Fig. 3. Growth (●) and changes in extracellular glucose (△), protein (□) and amylase (■) during fermentation with glucose as carbon source.

Acknowledgement

This work was supported by The Danish Natural Science Research Council.

REFERENCES

Adams, P. R., 1981, Amylase Production by Mucor Pusillus and Humicola Lanuginosa as Related to Mycelial Growth, <u>Mycopathologia</u>, 76:97.

Barnett, E. A., and Fergus, C. L., 1971, The Relation of Extracellular Amylase, Mycelium, and Time in Some Thermophilic and Mesophilic Species, <u>Mycopath. Mycol. Appl.</u>, 44:131.

Basaveswara Rao, V., Maheshwari, R., Sastry, N. V. S., and Subba Rao, P. V., 1979, A Thermostable Glucoamylase from the Thermophilic Fungus Thermomyces Lanuginosus, <u>Current Science</u>, 48(3):113.

Domsch, K. H., Gams, W., and Anderson, T-H., 1980, "Compendium of Soil Fungi", Academic Press, London.

Manning, G. B., and Campbell, L. L., 1961, Thermostable Alpha-Amylase of Bacillus Stearothermophilus, <u>J. Biol. Chem.</u>, 236:2952.

PART IV
CELLULOLYTIC ENZYMES

ANALYSIS OF MULTI-FORM CELLULASES OF

SPOROTRICHUM CELLULOPHILUM

Shinichi Kinoshita, Masahide Tamaki, Jung-Hoe Kim,
Ancharida Svarachorn, Alberto Araujo and Hisaharu Taguchi

I. C. Biotech, Faculty of Engineering
Osaka University
Suita, Osaka, Japan

ABSTRACT

Sporotrichum cellulophilum produces 2 β-glucosidases and about 10 cellulases. From the culture filtrate, 2 β-glucosidases and 5 cellulases were purified to homogeneity by DEAE-Sephadex A-50, Toyo-pearl TSK-HW55 chromatographies and preparative polyacrylamide gel electrophoresis. Among them cellulase CI was the smallest, CII was the most abundant, CIV had a very high activity on cellulose powder, and CV had a very high specific activity. The rabbit antiserum prepared against cellulase CIV reacted with CII and CV. The multi-form cellulases in the strain may be formed due to modification, and this organism may initially produce only a few cellulases.

INTRODUCTION

A new strain Sporotrichum cellulophilum was isolated from soil as a high cellulase producer. The cellulase activity was thermostable up to 60°C and hydrolyzed cellulose powder mostly to glucose. The enzyme was used to saccharify cellulose powder by using an ultrafilter membrane reactor [1].

Here we describe the purification of cellulases from S. cellulophilum, the existence of multi-enzyme components and their relationship.

METHODS

Analyses. Protein was determined from the absorbance at 280 nm and estimated by assuming that one A_{280} unit corresponded to 0.5 mg/ml.

Enzyme activity was determined by using 3 substrates at 40°C at pH 5. One unit of activity was defined as an amount producing 1 μmol of product in 1 min. An enzyme solution was reacted with 1% carboxymethyl cellulose (CMC) solution or 1% cellulose powder (KC-floc, Sanyo Kokusaku Pulp Co.) suspension, and the released reducing sugar was determined by the Somogyi-Nelson method. When p-nitrophenol-β-D-glucoside (PNPG) was used as a substrate, the released p-nitrophenol was determined by the absorbance at 420 nm.

Preparation of crude enzyme. The fungus was cultivated in a liquid medium for 3 days by using 50-1 fermenter. The culture broth was concentrated by ultrafiltration to one-tenth of its volume after removal of mycelia, and lyophilized. This enzyme powder (1 g) contained 0.26 g protein, and showed 120, 600, and 60 units of activity on PNPG, CMC, and KC-floc, respectively.

Preparation of antibody. The purified cellulase CIV (1 mg) was dissolved in 1 ml of 0.2% sodium dodecyl sulfate (SDS), boiled for 10 min, and emulsified with 1 ml of Freund complete adjuvant (Difco Co.) by sonication. It was subcutaneously injected on the back of two New Zealand white rabbits. After 2 weeks, 2 subcutaneous injections of the enzyme emulsion prepared with Freund incomplete adjuvant (Difco Co.) were carried out at an 8-day interval. One week after the last injection, the blood was withdrawn and the antiserum was obtained.

Preparative polyacrylamide gel electrophoresis. The gel is hollow cylinder-shaped with outer diameter 46 mm, inner diameter 20 mm, and 90 - 100 mm in height. Both sides of the gel were cooled and electrophoresis was usually carried out at pH 8 with a constant current of 25 mA after loading with 50 to 70 mg of protein. After about 16 h electrophoresis, the gel was taken out from the apparatus (Toyo Filter Co., model CD 50), sliced into rings 2.5 mm thick, broken by syringe passage.

RESULTS AND DISCUSSION

Purification of cellulases and β-glucosidases. The crude enzyme (20 g) was suspended in 100 ml of 50 mM phosphate buffer, pH 6.7, and the precipitate was removed by centrifugation. The supernatant was passed through a Sephadex G-25 column (ϕ 35 x 400 mm) to remove salts. The active fractions were charged on a DEAE-Sephadex A-50 column (ϕ 35 x 500 mm) and proteins were eluted with 2-1 linear NaCl gradient (0 to 1 M in the phosphate buffer). Most of the cellulase was eluted at 0.2 - 0.3 M NaCl together with β-glucosidase and a minor activity was eluted at 0.6 M NaCl, which was named cellulase CII. The major peak fractions were concentrated by ultrafiltration with the use of 20,000 molecular cut-off membrane, and charged on a Toyo-pearl TSK-HW55 (Toyo-Soda Co.) column for gel filtration. Its elution pattern is shown in Figure 1. β-Glucosidase activity appeared in two peaks, which were named β-glucosidase βI and βII.

Cellulase activity was detected in many fractions, which were named cellulase CII, CIII, CIV, and CV according to the elution order as indicated in the Figure. Among them cellulase CII was a main fraction, and CV had a rather high specific activity. All enzyme fractions were pooled, concentrated by ultrafiltration, and further purified by preparative polyacrylamide gel electrophoresis at pH 8. As an example the purification of cellulase CII is shown in Figure 2, where fraction number corresponds to a purified cellulase CII fraction. By carrying out the electrophoresis of other fractions, cellulase CI, CII, CIV and CV, and β-glucosidase βI and βII were purified to homogeneity, which are shown in Figure 3. The purification of each enzyme is summarized in Table 1.

The activities on CMC, KC-floc, and PNPG of purified enzymes were determined as shown in Table 2. β-Glucosidase βII had a little activity on CMC, but βI had none. Cellulase CV had a quite high activity on KC-floc, which was one-quarter of that on CMC, CII had some activity on PNPG.

Some properties of purified enzymes. The specific activities of β-glucosidases βI and βII were almost the same. Cellulase CI had a very low specific activity, CV had a very high specific activity.

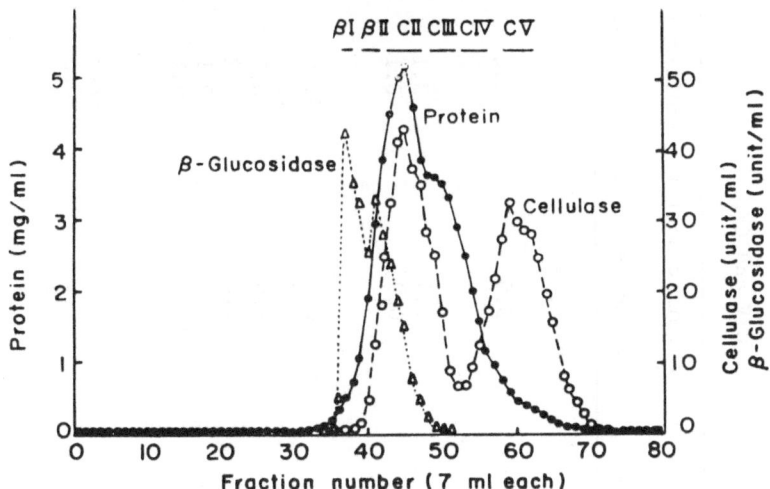

Fig. 1. Purification of cellulase and β-glucosidase by Toyo-pearl TSK-HW55
column chromatography. The active fractions in DEAE-Sephadex A-50
chromatography were charged after concentration. β1, β2, CII,
CIII, CIV and CV indicated in the figure were pooled and concentrated and further purified by preparative polyacrylamide gel
electrophoresis.

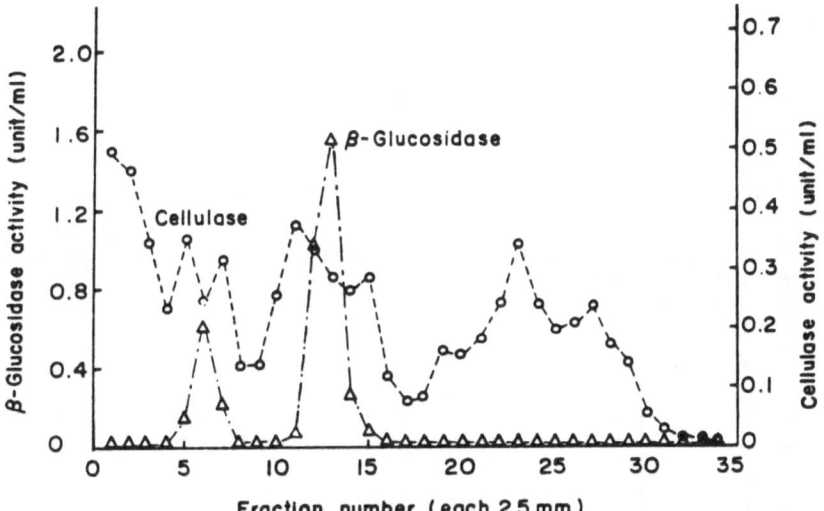

Fig. 2. Purification of cellulase CIV by preparative polyacrylamide
gel electrophoresis. The cellulase CIV fraction from HW-55
chromatography was charged. After electrophoresis at pH 8, the
enzyme was eluted as described in Methods and the activities on
CMC and PNPG in each fraction were determined.

The molecular weight was determined by Sephadex G-200 column (ϕ 25 x
1,000 mm) chromatography using marker proteins (Boehringer GmbH) and the
subunit molecular weight by SDS-polyacrylamide gel electrophoresis with
marker proteins (Boehringer GmbH and Seikagaku Kogyo Co.). The sugar
content was determined by the phenol sulfuric acid method.

Table 1. Summary of Purification of β-glucosidases βI and βII, Cellulases CI, CII, CIV and CV.

Purification step and fraction	Total protein (mg)	Total activity (unit)		Specific activity (unit/mg)	
		cellulase	β-glucosidase	cellulase	β-glucosidase
Crude enzyme	3 260	7 800	1 800	2.4	0.6
Sephadex-G25	1 280	7 600	1 700	5.9	1.4
DEAE-Sephadex A-50					
main peak	784	5 800	1 500	7.4	3.1
CI	132	120	4	0.9	0.1
Toyo-pearl TSK-HW55					
βI	13	10	650	0.7	49.0
βII	56	56	590	1.0	11.0
CII	93	610	65	6.6	0.7
CIII	69	190	16	2.8	0.2
CIV	59	150	6	2.6	0.1
CV	21	370	2	16.0	0.1
Preparative electrophoresis					
βI	1.2	0	236	0	197.0
βII	1.3	4	205	3.1	160.0
CI	16.7	32	0	1.9	0
CII	20.5	140	4	6.8	0.2
CIV	17.3	50	1	2.9	0.1
CV	4.6	123	0	27.0	0

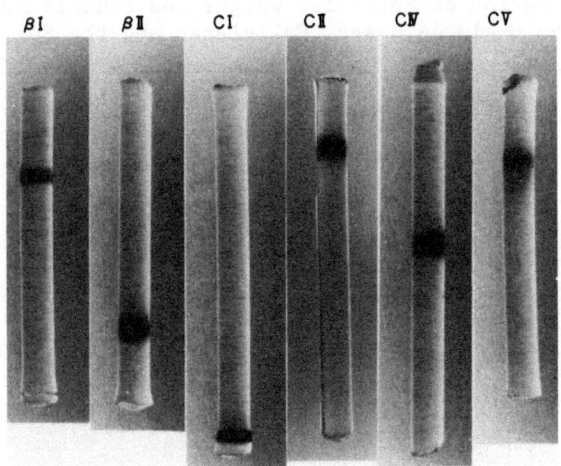

Fig. 3.   Polyacrylamide gel electrophoresis at pH 9 of purified
β-glucosidases βI and βII, cellulases CI, CII, CIV, and CV.

The molecular weight of purified enzymes was determined by Sephadex
G-200 column chromatography. The results are summarized in Table 3. The
largest enzyme was β-glucosidase βI (260,000), and the smallest was cellulase CI (39,000). The molecular weight of enzyme subunits was also determined by SDS-polyacrylamide gel electrophoresis as shown in Table 3. These
data showed discrepancy with the data of gel filtration. For example, the
molecular weights of cellulase CIV and CV were higher than those of each
subunit and the molecular weights of each enzyme were quite different from
integers of those of each subunit. To check this discrepancy, we determined the molecular weight of cellulase CIV by equilibrium centrifugation
(carried at the Inst. Protein Res., Osaka University) to be 70,000. We
thought this strange result may be caused by a glycoprotein nature of the
enzymes. The sugar moiety would cause the retarded elution in gel filtration by interacting with the gel, and therefore the molecular weight
would be underestimated. The sugar would also cause slow migration in
electrophoresis due to the interference, and therefore the subunit molecular weight would be overestimated.

We found that all the enzymes contained 6 to 12% sugar as shown in
Table 3. From these considerations cellulase CII would be a monomer
protein, whose molecular weight was about 70,000. We supposed that β-
glucossidase βI may be a tetramer, βII a dimer, and cellulases CI and CII a
dimer, CIV and CV monomers.

Table 2. Substrate Specificity of Purified β-glucosidases and Cellulases.

		Relative activity (%) on		
		CMC	KC-floc	PNPG
β-glucosidase	βI	0.7	0.4	100
	βII	3.7	0.2	100
Cellulase	CI	100	4.1	0.3
	CII	100	8.5	5.3
	CIV	100	24.0	0.5
	CV	100	4.5	0.2

Table 3. Determination of Molecular Weight and Subunit Molecular Weight of Purified β-glucosidases, Cellulases and the Sugar Content.

		Molecular weight			
		Sephadex G-200	SDS PAGE	Equilibrium centrifugation	Sugar content (%)
β-glucosidase	βI	260 000	110 000		14
	βII	98 000	72 000		20
Cellulase	CI	39 000	34 000		6
	CII	75 000	57 000		15
	CIV	52 000	95 000	70 000	12
	CV	45 000	67 000		25

Subsequent purifications yielded more proteins with cellulolytic activities besides the purified 6 proteins, and we are purifying other proteins. The existence of the multi-enzyme system was reported in cellulases of Trichoderma viride (T. reesei) [2], Aspergillus niger [3], Chrysosporium lignorum [4], Irpex lacteus [5], and Aspergillus aculeatus [6]. In view of the cell economy, fungal cell is not likely to synthesize so many kinds of proteins having the same activity. We suppose that the organism may produce one or two enzymes, and these are then modified later by proteolytic and glycosidation processes.

In order to clarify the modification steps of cellulases in S. cellulophilum, we prepared the antibody against purified cellulase CIV, which had a high activity on KC-floc. This rabbit antiserum could react with the enzyme at 4-fold dilution and with as little as 6×10^{-6} enzyme units in a immunodiffusion test. The CIV antibody reacted with the fractions 44 through 66 of TSK-HW55 chromatography in Figure 1, and fractions 5 through 13 of preparative polyacrylamide gel electrophoresis in Figure 2. The CIV antibody reacted with the purified cellulase CII and CV but not with purified cellulase CI and β-glucosidases βI and βII as shown in Figure 4. These results showed that cellulase CIV was related with cellulase CII and CV but not with cellulase CI and β-glucosidases βI and βII.

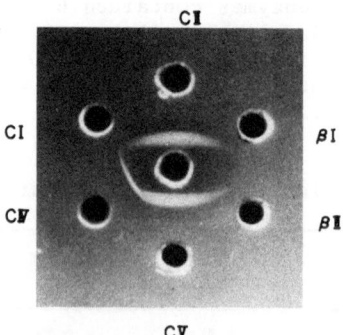

Fig. 4. Immunodiffusion test of rabbit antiserum prepared against cellulase CIV with purified enzymes. In center well 10 μl of undiluted antiserum was placed and the outer wells contained each 10 μl of β-glucosidase βI (1.5×10^{-3} unit) and βII (1.2×10^{-3} unit), cellulase CI (2.0×10^{-5} unit), CII (1.0×10^{-4} unit), CIV (5×10^{-5} unit), and CV (5×10^{-4} unit).

By using polyclonal antibody it is very difficult to show the detailed difference or relationship between the enzymes, and we are now trying to prepare monoclonal antibody by using the hybridoma technique. If we obtain many cell lines, we will show the relationship of cellulases more quantitatively. After this we will prepare monoclonal antibody by using a peptide fragment of cellulase, which will give a more specific reaction with other cellulases. Also we are now working on cloning of the S. cellulophilum cellulase gene in order to obtain unmodified cellulase.

Acknowledgement

We thank Mr Miwa of the Ajinomoto Co. for preparing the crude cellulase of S. cellulophilum and cellulase CIV antibody.

REFERENCES

1. S. Kinoshita, W. J. Chua, N. Kato, T. Yoshida, and H. Taguchi, _Enzyme Microb. Technol._, 8 (1986) in press.
2. S. Toda, H. Suzuki, and K. Nishizawa, _J. Ferment. Technol._, 49:499 (1971).
3. K. Ikeda, T. Yamamoto, and M. Funatsu, _Agric. Biol. Chem._, 37:1153 (1973).
4. K. E. Erikson, and B. Patterson, _Eur. J. Biochem._, 51:193 (1975).
5. T. Kanda, K. Wakabayashi, and K. Nishizawa, _J. Biochem._, 79:977 (1976).
6. S. Murao, J. Kanamoto, R. Sakamoto R., and M. Arai, _J. Ferment. Technol._, 57:157 (1979).

CELLULASE SECRETION FROM A HYPERCELLULOLYTIC MUTANT

OF TRICHODERMA REESEI RUT-C30

B. K. Ghosh, A. Ghosh and Adrian Salnar

UMDNJ - Robert Wood Johnson Medical School
(Formerly Rutgers Medical School)
Piscataway, N.J. 08854, USA

Cellulose is the most broadly distributed biomass of the world [1].
This biomass is a polymer [1] of glucose and is depolymerized by an enzyme
cellulase [2,3]. This depolymerization is continuously occurring in nature
to provide nutrients to herbivorous animals and other living organisms.
This enzyme is primarily produced by microorganisms associated with decay-
ing plant material or in the digestive system of the animals. There has
always been considerable interest in cellulose degradation because the
resulting products may be used in various ways. This use may be e.g., the
fermentation of glucose to alcohol for fuel production, the use of glucose
as food product, the use of partially degraded cellulose as noncaloric food
additive and absorbent material or binding material. Although cellulose
may be degraded by acids an enzymatic degradation appears to be preferable
because the processing technology will be cost effective due to simplicity,
but enough cellulolytic enzymes must be available at a low cost of production.

The cellulase is produced by a large number of microorganisms,
particularly fungi [2,3], but this production is regulated by catabolite
repression. Therefore, only small quantities of this enzyme are produced
and may be available unless some specific enhancement has been made by
genetic manipulation - by conventional mutation technology or the more
specific recombinant DNA technology. Utilizing the former technology a
hypercellulolytic mutant was selected from Trichoderma reesei strain QM6a.
This mutant strain Rut-C30 has been a model for our study of cellulase
secretion. By definition, cellulase is an enzyme which can hydrolyze
crystalline cellulose to glucose. Three basic types of enzymes are needed
to complete this process, i.e., endo-1,4-β-D-glucanase (EC 3.2.1.4), cellobiohydrolase (EC 3.2.1.91), and cellobiase (β-glucosidase, EC 3.2.1.21).
The action of these three enzymes appears to be sequential and synergystic.
Therefore, any hypercellulolytic organism must exhibit proportionate
enhancement of these three enzymes; furthermore these three enzymes must be
secreted into the extracellular medium. Our continuing studies have shown
that Rut C-30 [4] is hypercellulolytic by producing about three times more
activity to hydrolyze crystaline cellulose to glucose relative to the wild
type QM6a; however, endoglucanase production is much more enhanced (i.e.,
15-18 times) than total cellulase or β-glucosidase [5,6]. In addition to
hyperproduction almost the entire amount (>90%) of endoglucanase is
secreted. The most intriguing phenomenon is the enhancement of endoplasmic
reticulum (ER) during the endoglucanase synthesis. In order to understand

the mechanism of regulation of cellulase synthesis and secretion we further studied the subcellular processes. The data show that the secreting hyphae contained an abundance of distended ER and single lobulated Golgi. Endoglucanase is synthesized in the ER in a catalytically active form but moves to another vesicular fraction which appears to be a Golgi fraction, in an inactive form.

MATERIALS AND METHODS

Culture of T. reesei: The method for the growth of T. reesei has been described in earlier publications [6]. The cells were usually induced for cellulase production with 1% Avicel PH 101. The mycelial preparation from this growth medium is unsuitable for subcellular fractionation due to the firm attachment of large amounts of Avicel crystals to the mycelia. In modified growth conditions washed spores were grown for 18h; the germinated mycelia were washed and induced in the presence of 1.0% lactose and 0.05% Avicel PH 101. After 18 h of growth endoglucanase started to be synthesized at a fast rate and around 24 h of growth secretion started at a rapid rate; after 36 h the synthesis slowed down, but secretion continued at a rapid rate. The washed mycelial preparation from this medium had no attached cellulose crystals and hence was suitable for fractionation. The mycelia after growth were harvested by centrifugation at 16 000 x g for 20 min and washed by filtration on a coarse sintered-glass filter with 0.01 M Tris-hydrochloride buffer (pH 7.0) containing 0.15 M NaCl.

Cell fractionation: Details of the fractionation procedure for cytoplasmic material freed of nuclei and mitochondria has been described in detail in a previous publication [7]. However, a brief description is given in a chart (Figure 1). Figure 2 shows the densities of the sucrose in the linear gradient of each collected fraction. The fractions were collected by a peristaltic pump (Pharmacia Inc.) from the bottom of the tube containing subcellular particles.

Enzyme assay, radioactive labeling and immunoprecipitation: Endoglucanase, cytochrome-c-reductase, ATPase, alkaline phosphatase and protein assay methods have been described in a previous publication [7]. Radioactive labeling was done with ^{3}H-L amino acid mixture (Sigma). The labeled amino acids were added to the media at a specified growth period and the incubation continued for a period of 15 min. These labeled cells were harvested, washed and used for subcellular fractionation. Immunoprecipitation was done with the antibody against purified endoglucanase secreted from Rut C-30. The antibody was prepared in rabbit as described earlier. The subcellular fractions were assayed for endoglucanase activities and the endoglucanase protein present in these fractions was determined by comparing it with the specific activity of the purified enzyme. The samples containing 0.1μg, 0.3μg, 0.5μg and 1.0μg of endoglucanase proteins were diluted to bring the volume to 400μl; 20μg of endoglucanase antibody in a volume of 100μl was added to these samples for precipitation by 60 min incubation at 37°C followed by overnight incubation at room temperature. The immunoprecipitates were collected by centrifugation at 3 000 x g for 30 min; the supernatant was removed by aspiration and the precipitates were washed with phosphate buffer followed by deionized water. Finally, these precipitates were dissolved in 10ml aquasol and the radioactivity present was counted in a scintillation counter fitted with a ^{3}H window. The controls were identical samples without the addition of antibody.

Ultrastructural examination: The mycelial samples were collected from the culture at different days of growth in the presence of different substrates, namely 1% Avicel PH 101, lactose, glycerol, glucose and a mixture of 1% Avicel and glycerol. These cells were processed as described earlier

Subcellular Fractionation of ^{3}H-labelled *T. reesei* Rut-C30

Mycellia germinated from spores in 1% glucose medium were washed free of glucose and grown in 1% lactose + 0.05% avicel (pH 101) for 18 or 24 hours.

↓

^{3}H-L-amino acid mixture was added to these mycellia and further grown for 15 minutes

↓

These mycellia were washed free of radioactivity and suspended in 0.01M Tris-HCl buffer, pH 7.0, containing 0.15M NaCl and 3mM $MgCl_2$.

↓

Mycellia were ruptured in a Bead Beater for two 3 min stretches; the process was pulsed at 15 second intervals.

↓

This broken cellular material was centrifuged at 500 × g for 10 min to remove unbroken cells.

↓

The supernatant was centrifuged at 5000 × g for 10 min to remove nuclei and mitochondria.

↓

The supernatant was layered on a continuous sucrose gradient (density 1.06 to 1.19 or 14 to 42%) and centrifuged at 25,500 rpm for 2 hrs.

↓

Three ml fractions were collected and were diluted (1:1) with the above buffer.

↓

The fractions were centrifuged for 2 hours at 80,000 × g to sediment the particulate matter. The above buffer (0.5 ml) was added to each particulate residue and the material was dispersed. These were used for the following:

Biochemical analysis after light sonication:

1) Endoglucanase
2) Mg-dependent ATPase
3) Alkaline phosphatase
4) Cytochrome-C-Reductase
5) Protein
6) Amount of total radioactivity
7) Amount of anti-endoglucanase IgG precipitated radioactivity

Ultrastructural analysis:

Fractions were fixed in a mixture of tannic acid-gluter-aldehyde at 0° followed by OSO_4. The thin sections were stained to examine in an electron microscope.

Fig. 1. A brief description of the methods of fractionation of T. reesei subcellular material.

in glutaraldehyde-tannic acid followed by OsO_4. The thin sections were examined in a JEM 100C (JEOL, Japan) electron microscope. Stereological analysis was done following a procedure described earlier to estimate the ER and other subcellular components.

RESULTS

Growth and endoglucanase production: Growth of the strains QM6a, the wild type, and the mutant Rut-C30 has been compared. It is evident from Figure 3 that mycelial growth of QM6a stops after 36-48 h of incubation; after this period the mycelial mass remains constant up to 120 h - then it

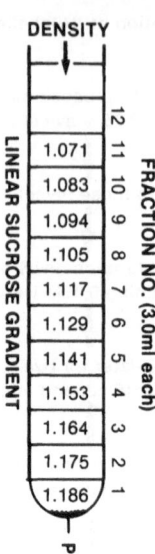

Fig. 2. Densities of the sucrose solution present in the subcellular fractions separated in the gradient.

slowly declines to 20-25% of the maximum. The Rut-C30 mycelia, on the other hand, continue growing for 216 h (8 days); this growth proceeds at a fast rate for 72 h.

Endoglucanase synthesis: The synthesis in the whole culture of Rut-C30 was about 15 times the amount synthesized by QM6a (Figures 4a,b,c). The synthesis progressed at a very fast rate in the early phase of the growth, reaching 90% of the total activity within 48 h of growth. The mycelia-bound activity increased at a fast rate, reaching 90% of the total within 36 h; however, significant amount of this activity was lost and mycelia retained only 25-30% of the total activity. About 90% of the endoglucanase activity was secreted, but at the early stages of growth the rate of secretion was slow, resulting in the accumulation of the enzyme in the mycelia. In QM6a an insignificant amount of the enzyme was secreted.

Derepression of endoplasmic reticulum: In our previous publication it has been demonstrated that the ER of Rut-C30 is enhanced in contrast to the QM6a mycelia while grown in Avicel. This study was further continued to examine whether this is repressed in the presence of glycerol. Data presented in Figure 5 show that proliferation of the ER is fully repressed if glycerol is present in the growth medium simultaneously with cellulose (Avicel). The synthesis of endoglucanase is also repressed by glycerol. However, the amount of ER is much lower if the mycelia grow in the presence of glycerol or glucose alone than with both Avicel and glycerol. Lactose is a poor inducer for endoglucanase, ER proliferation is also insignificant.

Ultrastructure of the ER and Golgi: The quantitative estimates presented above showed that the mycelia of Rut-C30 had extensively proliferated ER when grown in the presence of cellulose. This proliferation did not follow a pattern and exhibited extensive polymorphism. A few examples are shown in Figure 6. In Figure 6a, linear segments of the ER form multiple layers with parallel arrangement. The ribosomes are closely packed on the membrane surface. The cisternal space is narrow but in the distal locations the space is distended. In Figure 6b the arrangement of the ER segments is random and resembles a network. The cisternal space is narrow

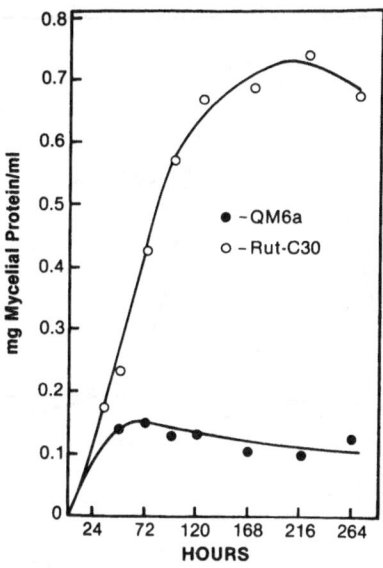

Fig. 3. Comparison of growth of two strains, wild type QM6a and Rut-C30, in the presence of 1% Avicel (pH 101); the mycelial mass of mutant Rut-C30 increases about 6 times more than QM6a.

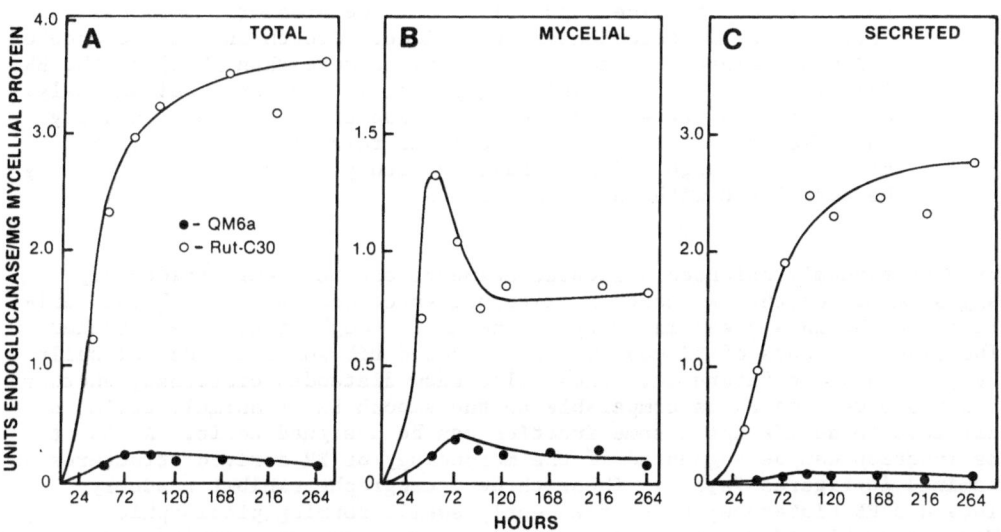

Fig. 4. Comparison of the synthesis of endoglucanase in QM6a and Rut-C30. (A) In the whole culture synthesis by Rut-C30 is about 15 times higher than QM6a; it proceeds at a fast rate, reaching 90% of the final amount within 48 h. (B) Mycelia-bound enzyme increases at a fast rate, reaching the maximum level within 36 h. Following this 50% of this bound activity is lost. (C) Overall about 90% of the endoglucanase activity is secreted; the rate of secretion proceeds much more slowly than the rate of synthesis, particularly in early stages, resulting in a rapid accumulation of the enzyme within the mycelia. A very small amount of the endoglucanase is secreted by QM6a.

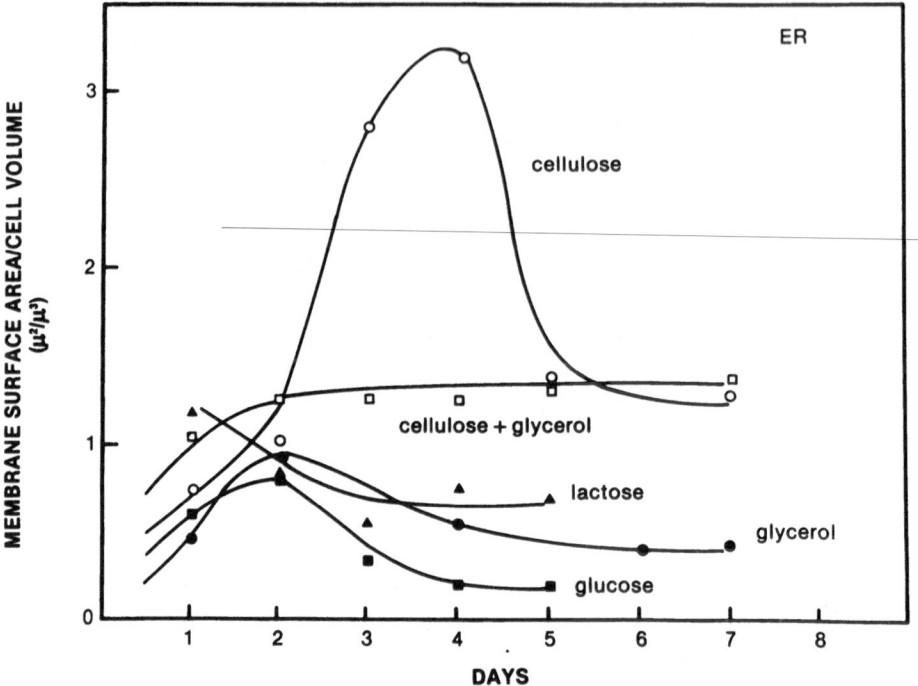

Fig. 5. Induction of endoplasmic reticulum (ER) in Rut-C30 by growth in the presence of Avicel (pH 101). Surface area of the ER was estimated by a stereological technique. Growth in the presence of cellulose causes an increase (2-3 times more than QM6a) of the ER. This increase is repressed if glycerol is present simultaneously. Growth in the presence of any other substrate does not show any increase of ER. As this organelle is related to secretory activity the high endoglucanase secretory activity of Rut-C30 may be correlated with this enhanced ER.

but many randomly oriented distended segments can be seen. Frequently, many segments of the ER showed a cisternal space continuous with the periplasm. Although a great majority of the ER is rough, i.e. have attached ribosomes, segments of ER mostly free of bound ribosomes are occasionally seen, as shown in Figure 6c. These also show distended cisternae; whether this ribosome free ER is comparable to the smooth ER of animals cells is difficult to decide until some function can be assigned to it. A clue to its function can be suggested if the morphology of ER derived structures is examined in Figures 7a,b,c. These three micrographs exhibit randomly distended ER cisternae; these frequently swell, forming pleomorphic bodies (Figures 7a,b). Some of these distentions are symmetric, giving a lobulated appearance. Careful examination of the membrane surface will show that these are largely free of ribosomes. It appears that secretory products accumulate inside the ER cisternal space and ER develops into the pleomorphic secretory granules by losing ribosomes from the membrane surface.

Examination of the Figures 8 and 9 will show the organization of the Golgi bodies. We reported earlier that Rut-C30 in spite of its hypersecretory activity does not show any characteristic Golgi. The use of glutaraldehyde-tannic acid fixative improved the definition of the membrane delimiting the saccules and their contents. Careful examination of the micrographs revealed the distribution and morphology of the Golgi system in

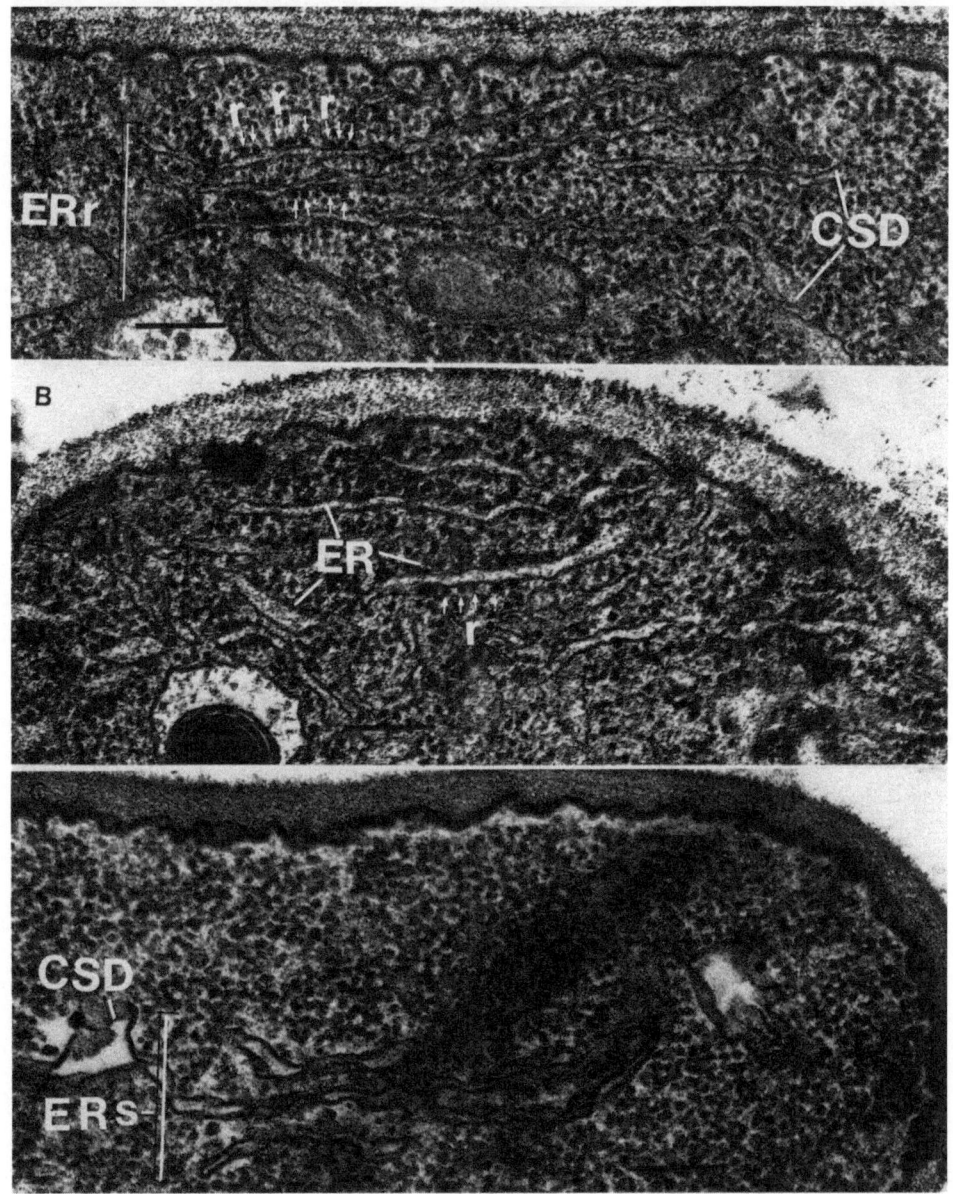

Fig. 6.   (A) The ER has a characteristic appearance; rows of ribosomes are attached to the membrane surface, the cisternal space is narrow but distinct and the distal ends are distended relative to the central region.   Extensive polymorphism is observed in the organization of the ER.   (B) Large number of short segments of ER may also be present and form a network.   The cisternae of these segments of ER may show irregular and asymmetric swelling with only few attached ribosomes.   Such randomly distributed bodies are present in large numbers in the hyphae of the secreting culture.   (C) In secreting hyphae smooth surfaced ribosome-free ER is frequently seen; the cisternae have distended ends.

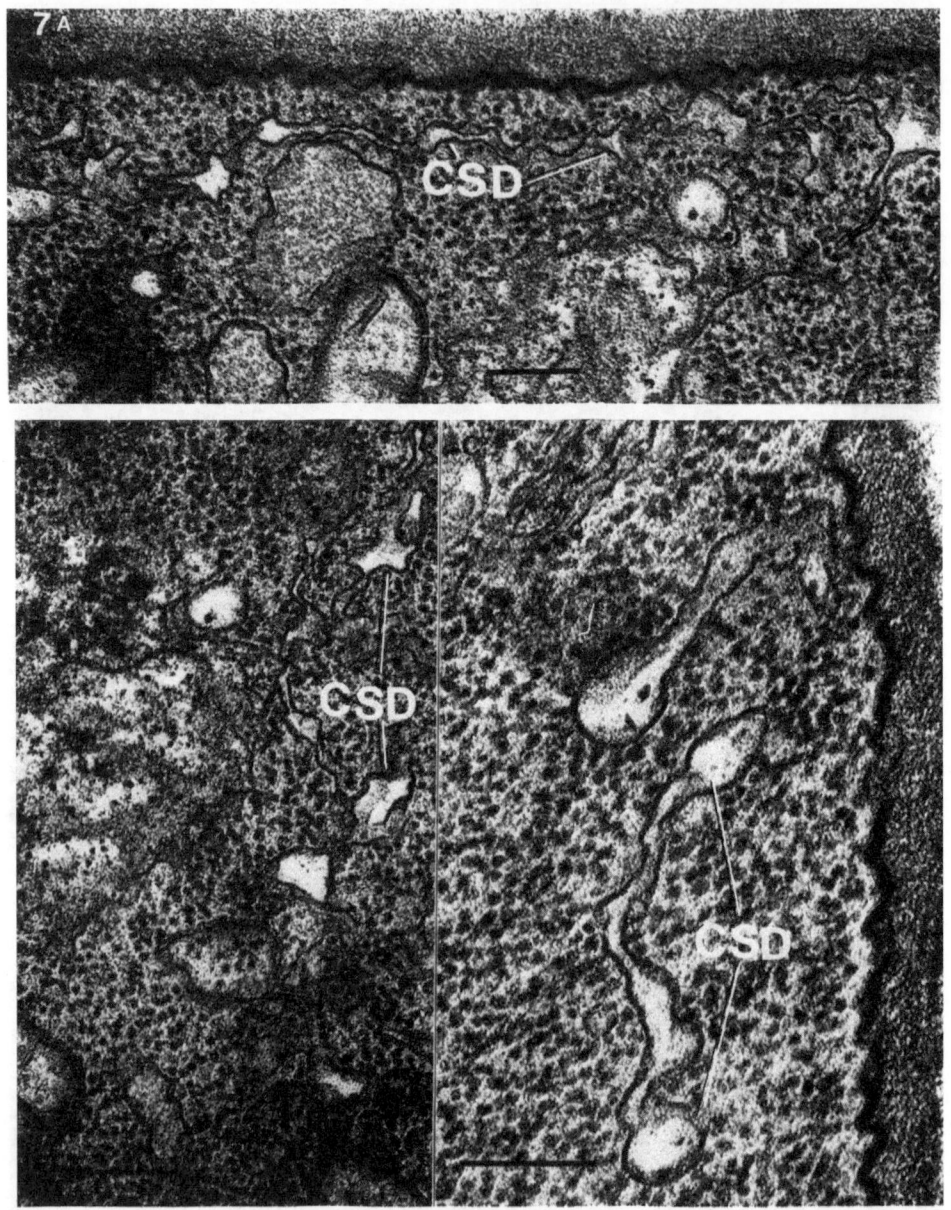

Fig. 7. (A) and (B) In rapidly secreting hyphae the ER is irregularly
distended and finally forms pleomorphic bodies filled with
secretory material. (C) The membrane surface at this stage is
ribosome free; this is evident from comparing membrane surfaces
of distended and undistended regions of the membrane.

Rut-C30 mycelia. As seen in Figures 8 and 9 the Golgi apparatus is concentrated, comparable to the animal cells in the perinuclear locations. In
both these figures the Golgi apparatus appears to by polymorphic, i.e. it
comprises randomly distributed vesicles or isolated single saccules, which
due to orientation may appear like networks (Figure 9). The vesicles
frequently form a collection in the vicinity of saccules (Figure 8). The

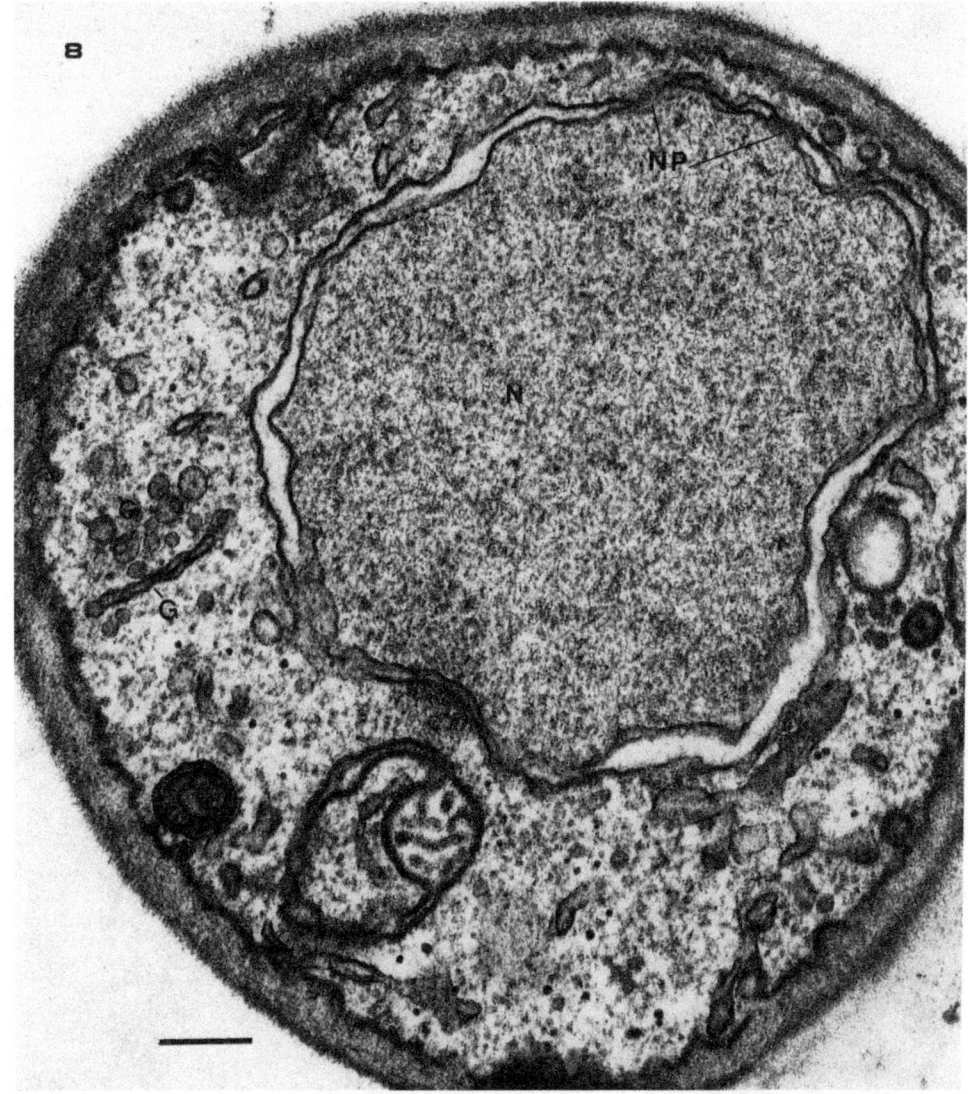

Fig. 8. Golgi apparatus is abundant but the bodies are vesiculated.

cisternal space is usually filled with dense material, but in spite of the
filling the saccules do not lose their shape (Figure 9).

Endoglucanase distribution in cell fractions: It is shown in Figure 2
that 12 fractions were collected. Fraction 1 is the heaviest and 12 is the
lightest; however, assays were also done with the precipitate. The distribution of the enzyme assays are given in Figures 10a-e; from the smoothness
of the distribution pattern of the enzymes in different cell fractions it
appears that the enzymes are in fact associated with specific subcellular
particles. As the fractions are on a continuous gradient there is a
significant amount of overlap. Therefore, if the fractions containing peak
values are compared an overall idea can be made regarding their subcellular
location. These can be described as endoglucanase 6, Mg-dependent-ATPase
6, cytochrome-c-reductase 3, Alkaline phosphatase 4 and the protein distri-

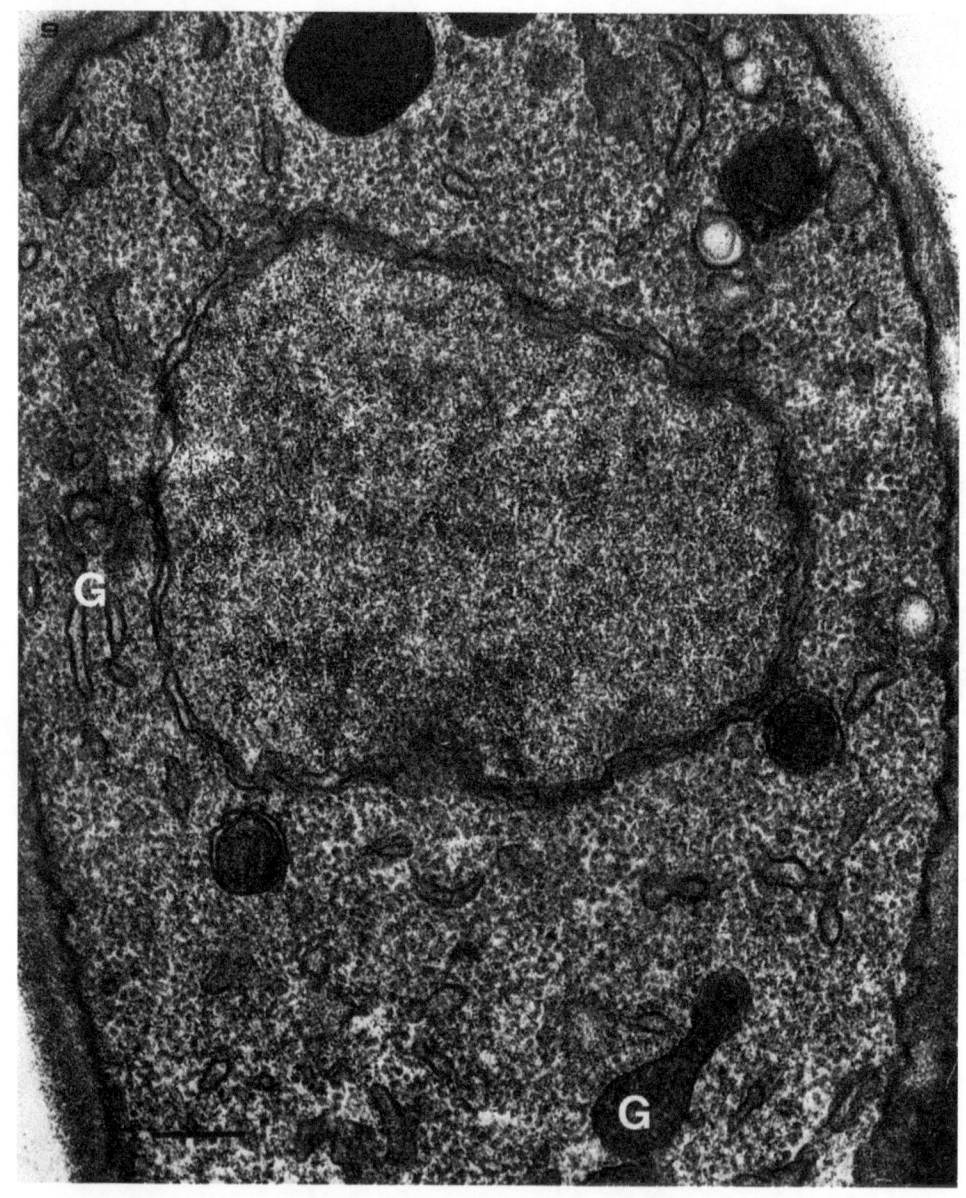

Fig. 9.   Isolated randomly distributed Golgi saccules which may be filled
          with secretory product. Perinuclear concentration of Golgi bodies
          is common.

bution remains comparable in most of the gradients. The precipitate fraction is poor in all the enzymes but the protein content is high. The
fraction 6 contains a large number of vesicles with ribosomes attached on
the outside surface of the membrane; this is a general characteristic of
microsomes derived from rough ER (Figures 11a,b). It is known that rough
ER is the site of extracellular enzyme synthesis. The vesicles of this
fraction appear largely to be empty because they are electron-lucid.
However, a few vesicles are seen to be filled with electron dense material
having crystalline structure (Figure 11c). The resolution in this gradient

was improved, compared to our previous report, and it appears that endo-glucanase and ATPase are associated with the light microsomal fraction whereas cytochrome-c-reductase and alkaline phosphatase with the heavy vesicle fraction. A complete ultrastructural identification is under way and it seems from the preliminary results that cytochrome-c-reductase concentrates in the Golgi rich fraction. However, this comment requires

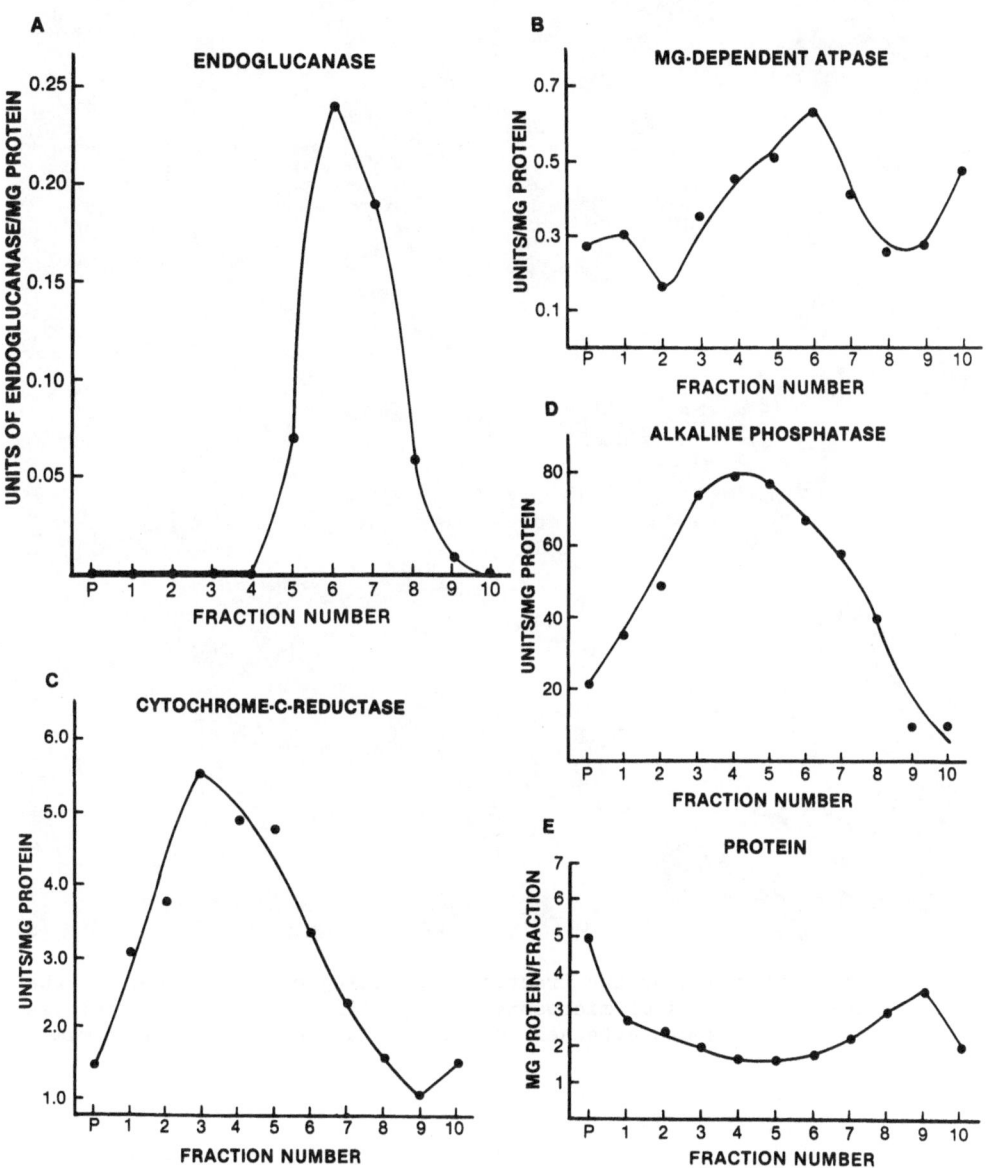

Fig. 10. Subcellular fractions prepared from an actively synthesizing population of hyphae shows the following distribution. (A) Endoglucanase is present in the light microsomal fraction. (B) Mg-dependent ATPase comigrates with the endoglucanase fraction. (C) Cytochrome-c-reductase migrates in the heavy vesicle fraction. (D) Alkaline phosphatase migrates between (B) and (C). (E) Protein distribution remains almost constant throughout the gradient.

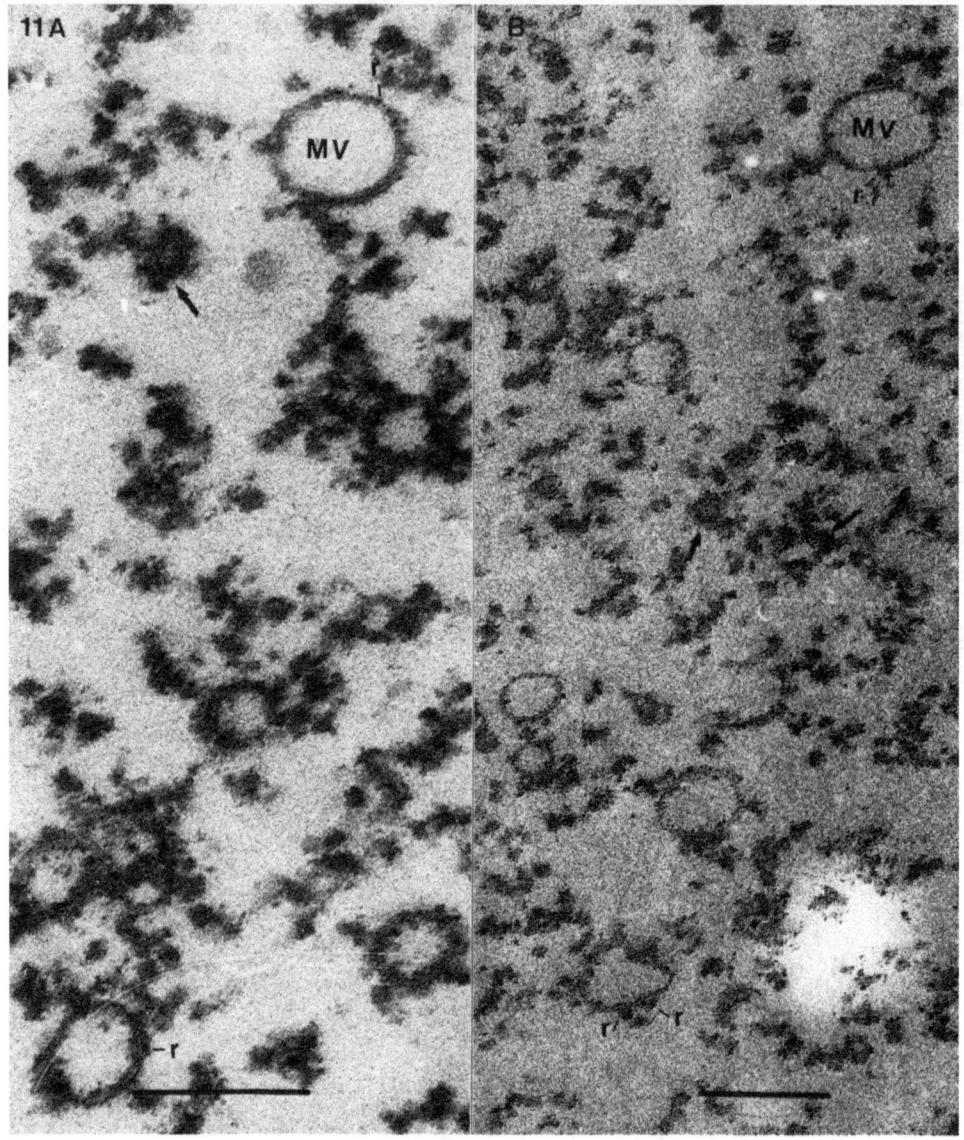

Fig. 11. Ultrastructure of the fraction containing endoglucanase has the
characteristics of microsome (A,B), different sizes of vesicles
having ribosomes attached to the outside surface of their mem-

further confirmation by a study using a high resolution gradient and extensive marker enzyme analysis.

In our previous paper we proposed that the endoglucanase may be synthesized in light microsomal vesicles and then transferred to a heavy
vesicle fraction in an inactive state. This issue has been examined in
experiments presented in this paper. The fractions were prepared from
mycelia at two different time intervals of growth, at 18 h when synthesis
was high and secretion was low and at 24 h when the reverse was true. The

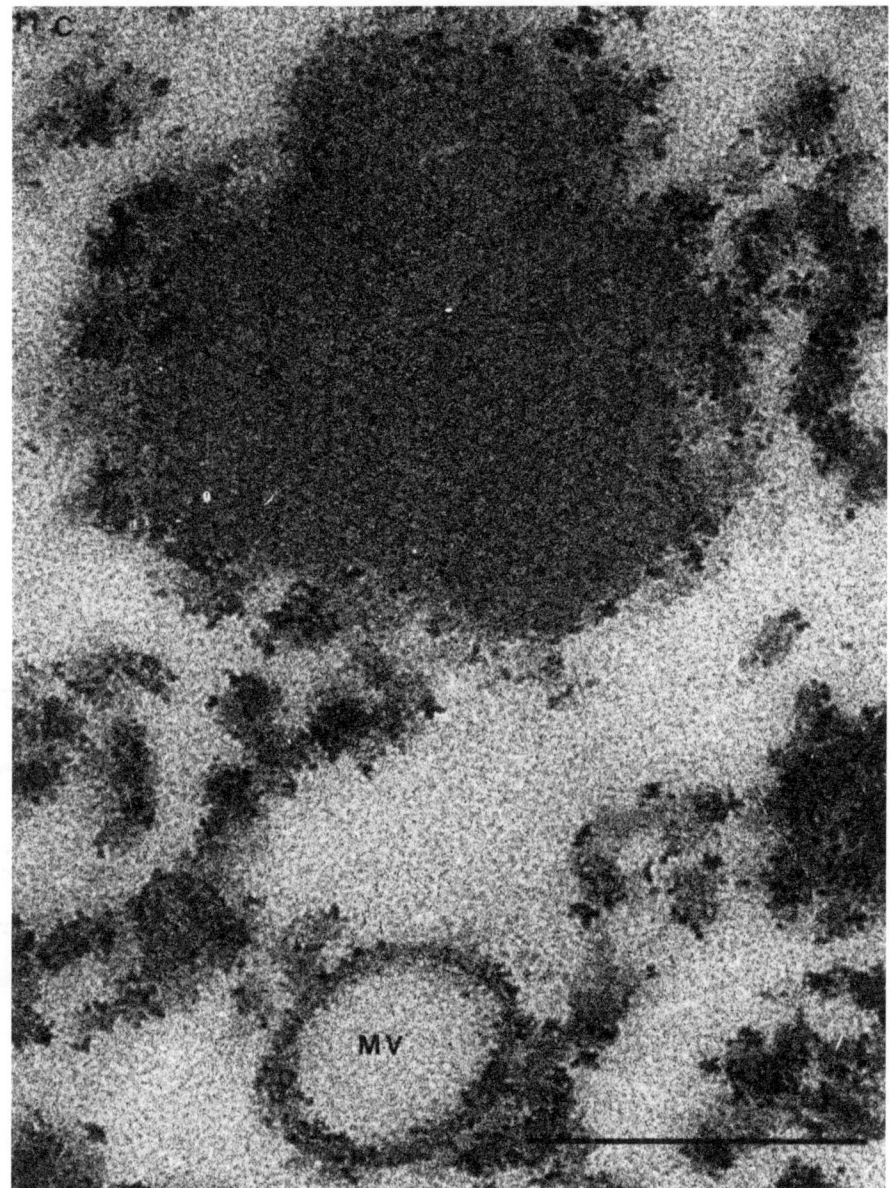

Fig. 11.   brane contain varying amounts of electron-dense material (arrow);
(cont..)   occasionally these vesicles contain crystalline material with
           two-fold symmetry (C).

results presented in Figure 12 show that at 18 h both the endoglucanase
catalytic activity and radioactively labeled immunoprecipitated endoglucanase concentrates in the light microsomal fraction. On the other hand, at
24 h the catalytic activity concentrates in the light vesicles and the
labeled immunoprecipitated material in the heavy vesicles fraction. These
results confirm our previous data and strongly suggest that during high
secretory activity an inactivation follows the synthesis of the endoglucanase.

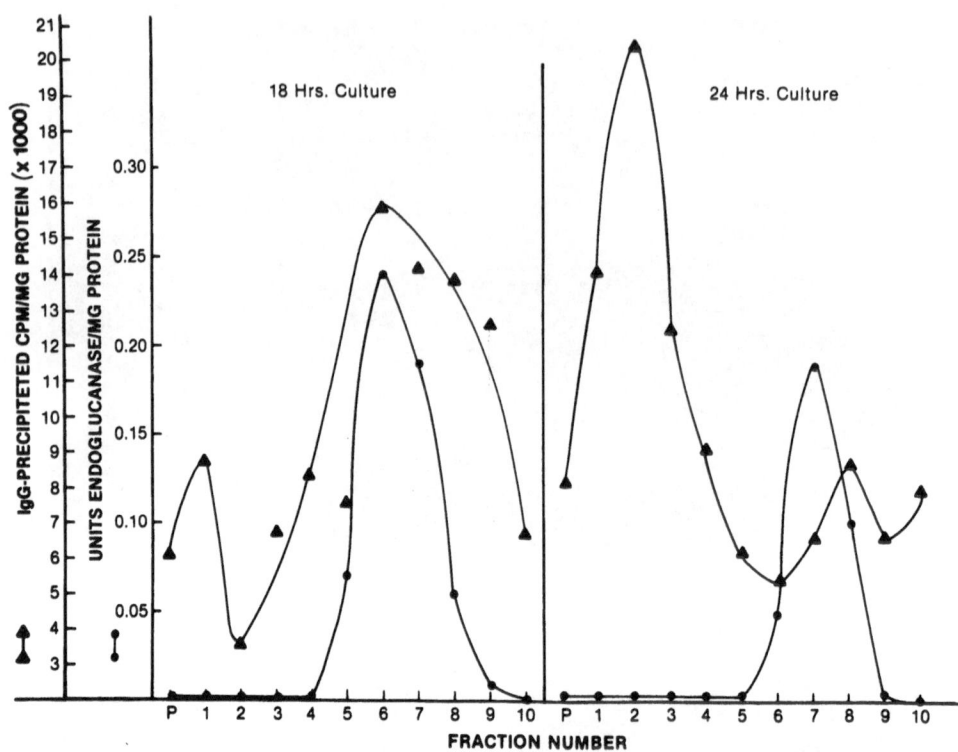

Fig. 12. Distribution of incorporated radioactivities in the endoglucanase antibody precipitated proteins are compared when the hyphal culture secretes only a small amount of endoglucanase (18 h) and, when it secretes at a high rate (24 h). (A) In a nonsecreting culture the endoglucanase activity and the immunoprecipitated radioactivity comigrates in the light microsomal fraction. (B) In a secreting culture endoglucanase activity and a minor amount of immunoprecipitated radioactivity are in the light microsomal fraction but the major amount of immunoprecipitated material accumulates in the heavy fraction.

DISCUSSION

The goal of the present study is to understand the mechanism and regulation of cellulase secretion from T. reesei. A hypercellulolytic mutant Rut-C30 has been used as a model. It has been shown in this paper and earlier that not only a considerably higher amount of endoglucanase is synthesized and secreted by this mutant but also that this occurs at a fast rate soon after germination of the spore. Therefore it is possible to conduct experiments with subcellular fractions from mycelia which are at a uniformly growing vegetative stage of growth. The knowledge of the subcellular pathway of protein (enzyme) secretion is extensive in animal cells and it is known that a multicomponent activity is involved [8,9]. Work on yeast cells indicates that the process may not be different in fungi [10]. However, in filamentous fungi the subcellular process of secretion is less well understood and may be considerably different particularly when the slow-growing mycelia adapt to the hypersecretion of a specific enzyme, in this case endoglucanase. The issue is how the mycelium commits itself to the synthesis and secretion of a high amount of a specific protein without jeopardizing its normal growth process. The usual secretory route, the growing tip of the mycelium, may not accommodate the outpouring of the specific protein. The clue to the hypersecretory activity in Rut-C30 lies in the proliferation of the ER and presumably a chain of secretory organelles. It is possible that repression/derepression of secretory proteins may be closely linked to the repression/depression of ER. The mutagenesis in Rut-C30 may have deregulated this specific process and thus hyperproduction and hypersecretion of endoglucanase can occur. Comparable deregulation may be a necessary condition for constructing strains of filamentous fungi capable of hyper-producing any secretory protein. Correlation of the repression of ER proliferation and endoglucanase synthesis is a very significant observation. It was believed that the organism is catabolite repression resistant; in fact endoglucanase synthesis is strongly repressed by glycerol in Rut-C30 strain. Therefore the repression of both ER and endoglucanase synthesis might not have been a mere coincidence. However, the genetic aspect of this regulation is complex; it is the fortuitous deregulation of ER proliferation which might have led to the increased synthesis and secretion of endoglucanase.

In addition to the changes in the ER during endoglucanase hypersecretion a large number of Golgi bodies were observed. These are distended with electron-dense contents and are randomly distributed particularly in the perinuclear region. The ER shown a significant polymorphic character in secreting hyphae. Some of the varying morphology appears to be related to the secretory activity, viz. the pleomorphic vesicular bodies formed by distention of the ER cisternal space; ribosomes dissociate from the membrane surface of these distended ER. These are present in abundance and frequently found to be associated with the plasma membrane.

Localization of endoglucanase in the microsomes and rapid incorporation of radioactivity into this endoglucanase indicate the site of synthesis in rough ER. This was found when the secretory activity was very poor; but during a high secretory activity the endoglucanase incorporating radioactivity was found in a heavy vesicle fraction and the catalytic activity of this enzyme was poor. This heavy vesicle fraction was found to be enriched with Golgi material. It is possible that this inactivation and crystallization constitute a part of the presecretory processing which occurs in the Golgi system. The enzyme was possibly synthesized in a catalytically active form; however, when the processing is slow the material is retained in the microsomal fraction. It has been shown that extensive processing and multiplicity of cellulases occurs at the secretion stage rather than at the post-secretional stage [11,12]. It is also possible that during the rapid secretion of the enzyme multiple processes of secretion are initiated leading to multiple molecular forms of the enzyme, and some of these may bypass the complex processing route to be secreted at a fast rate. However, these possibilities have to be sorted out by future experiments involving determination of the kinetics of movement of the enzymes in different subcellular compartments, characterization of the molecular forms of the enzyme in these compartments and examination of the turnover of the enzyme from these compartments to the extracellular medium. It seems that Rut-C30 is primarily a secretion mutant and thus enhancement of the cascade of events for secretion may stimulate the synthesis of the extracellular endoglucanase. The microbial cells may reserve a considerable amount of the latent capacity for secretory activity which might be linked by specific mechanisms to the synthesis of various secretory proteins. Thus enhanced secretion may be specifically exploited by an enhanced secretory protein synthesis, in this case endoglucanase synthesis.

Acknowledgement

This investigation was supported by National Science Foundation grant PCM 8407159. We thank Selma Topper for her help in preparing this manuscript.

REFERENCES

1. R. Malcolm Brown, Jr., Cellulose and Other Natural Polymer System - Biogenesis, Structure and Degradation, Plenum Press, New York and London, (1982).

2. R. D. Brown, Jr. and L. Jurasek (eds.), Hydrolysis of Cellulose: Mechanism of Enzymatic and Acid Catalysis, Amer. Chem. Soc. Adv. Chem. Ser., 181 (1979).

3. J. A. Gascoigne and M. M. Gascoigne, Biological Degradation of Cellulose, Butterworthsand Co., London (1960).

4. D. E. Eveleigh and B. S. Montenecourt, Increasing yields of extracellular enzymes, Adv. Appl. Microbiol., 25:57 (1979).

5. A. Ghosh, S. Al-Rabiai, B. K. Ghosh, H. Trimino-Vasquez, D. E. Eveleigh, and B. S. Montenecourt, Increased endoplasmic reticulum content of a mutant of Trichoderma reesei (Rut-C30) in relation to cellulase synthesis, Enzyme Microb. Technol., 4:110 (1982).

6. A. Ghosh, B. K. Ghosh, H. Trimino-Vazquez, D. E. Eveleigh, and B. S. Montenecourt, Cellulase secretion from a hyper-cellulolytic mutant of Trichoderma reesei Rut-C30, Arch. Microbiol., 140:126 (1984).

7. M. Glenn, A. Ghosh, and B. K. Ghosh, Subcellular fractionation of a hyper-cellulolytic mutant, Trichoderma reesei Rut-C30: localization of endoglucanase in microsomal fraction, Appl.Enrivonmental Microbiol., 50:1137 (1985).

8. J. D. Morre and L. Ovtract, Dynamics of Golgi apparatus: Membrane differentiation and membrane flow, Int. Rev. Cytol., (Suppl.):161 (1977).

9. G. E. Palade, Intracellular aspects of the process of protein synthesis, Science, 189:347 (1975).

10. P. Novick, S. Ferro, and R. Schekman, Order of events in yeast secretory pathway, Cell, 25:461 (1981).

11. H. Kolar, M. Mischak, W. P. Kammel, and C. P. Kubicek, Carboxy methylcellulase and β-glucosidase secretion by protoplasts of Trichoderma reesei, J. Gen. Microbiol., 131:1339 (1985).

12. W. P. Kammel and C. P. Kubicek, Absence of post secretional modification in extracellular proteins of Trichoderma reesei during growth on cellulose, J. Appl. Biochem., 7:138 (1985).

PRODUCTION OF β-GLUCOSIDASE IN <u>ASPERGILLUS TERREUS</u> ATG-5

O. Volfová and E. Kyslíková

Institute of Microbiology
Czechoslovak Academy of Sciences
Prague, Czechoslovakia

β-Glucosidase is an important component of the complex of cellulolytic enzymes converting cellobiose, a repressor of endo- and exo-β-glucanase, to the final product – glucose. Trichoderma reesei, one of the best producers of cellulases used in a pilot plant production of these enzymes, produces β-glucosidase exhibiting a low activity and this low activity is a limiting factor of hydrolysis of the cellulose substrate. For the commercial utilization of cellulolytic enzymes it is thus necessary to overcome some problems concerned with β-glucosidase (Woodward and Wiseman, 1982). These involve: 1. A low activity of extracellular β-glucosidase in culture filtrates of T. viride or T. reesei. 2. Inhibition of β-glucosidase by glucose, the final product of cellulolysis. The accumulated cellobiose then strongly inhibits endo- and exo-β-glucanases and thus the cellulolysis as a whole. 3. Heat inactivation of β-glucosidase during hydrolysis of cellulose. These problems can be solved in several ways such as: 1. Isolation of T. reesei producing higher amounts of β-glucosidase, 2. Selection and isolation of other producers with a high production of β-glucosidase. Their β-glucosidase would then be used in mixed enzyme preparations or, in its immobilized form, it would degrade cellobiose in cellulose hydrolyzates and thus increase the rate of hydrolysis. 3. Conversion of cellulolytic products to fructose which is only a weak inhibitor of β-glucosidase in T. reesei mutants. 4. Selection and isolation of producers forming β-glucosidase with altered properties, i.e. with an increased thermostability or decreased sensitivity of β-glucosidase against inhibition by the final product – glucose.

It follows from literature data that the fungi Botryodiplodia theobromae, Lenzites trabea, Sclerotium rolfsii and the genus Aspergillus, A. foetidus, A. fumigatus, A. niger and A. terreus in particular, are potential β-glucosidase producers. In the genus Aspergillus, as compared with Trichoderma, the rate of cellulose hydrolysis is limited by a low level of exo- and endoglucanases in the cellulolytic complex rather than by a low level of β-glucosidase.

A wild strain Aspergillus terreus isolated from wood drifted ashore in India (Araujo and D'Souza, 1980) exhibits an increased production of β-glucosidase as compared with other known producers of cellulases. Therefore, in our studies of cellulases we concentrated on this organism. We first determined the effect of cultivation conditions on the properties and production of individual enzymes of the cellulolytic complex (D'Souza and

Volfová, 1982). After determining the optimal cultivation conditions and optimal conditions for the enzyme production we started studies aimed at improving the wild strain of A. terreus.

The wild strain of A. terreus was improved by UV radiation and a subsequent selection of cellulolytic strains in the presence of a catabolite repressor of cellulases. Suspension of conidia of the fungus was irradiated and the irradiated conidia were incubated on an agar medium containing 0.5% pretreated crystalline cellulose (Tansey, 1971), 1.5% oxgall restricting the colony size and 5% glycerol at 30°C. After a 30-s irradiation about 6% and 4% of conidia survived in the medium without and with glycerol, respectively. After a subsequent incubation at 50°C transparent zones are formed around desirable mutant colonies. These colonies produce cellulases to the medium even in the presence of glycerol.

Of about 5 500 colonies tested on cellulose with glycerol we obtained 6 mutants with a cellulolytic activity manifested by transparent plaques on agar plates. These mutants designated ATG were further tested with respect to the production of the cellulolytic complex in cultivation flasks containing a mineral medium with 1% crystalline cellulose. An increased production of the cellulolytic enzyme complex was detected only in 3 mutants, i.e. in 50%. Detailed studies under constant cultivation conditions in a laboratory fermentor showed that the strain ATG-5 is the best producer of β-glucosidase. In this strain we studied the production and properties of the complete cellulolytic complex with a special attention to β-glucosidase.

The production of extracellular and bound β-glucosidase in ATG-5 under constant cultivation conditions (30°C, pH 5.0) in laboratory fermentor is illustrated in Figure 1c. The bound activity was determined from the difference between the activity of a culture suspension and the activity of culture filtrate considering also the sample volume correction. The onset of the production of all components of the enzyme complex is associated with the onset of cell growth. As compared with glucanases, the rate of the production of extracellular β-glucosidase is higher in the stationary phase of growth, almost linear during the whole cultivation time in some cases. The almost constant level of bound β-glucosidase in the stationary phase of the cell growth indicates that β-glucosidase is released primarily to the medium during this phase. The maximal activity levels of the whole cellulolytic complex are reached as late as during the post-stationary phase, when A. terreus ATG-5 produces 5 IU/mL and 20 IU/mL of extracellular β-glucosidase and endoglucanase, respectively.

The supernatant of the cultivation medium, after the cultivation of A. terreus ATG-5 in the mineral medium containing 1% cellulose, was used for the characterization of β-glucosidase.

It follows from Figure 2 that the optimal assay temperature for β-glucosidase is 70°C, as compared with that for exo- and endoglucanases. This property is important with respect to its possible application and has not yet been described in A. terreus. The temperature optimum is the same, both in aryl-β-glucosidase (measured with PNPG) and in cellobiase. In addition, in thermotolerant strains including our strain (maximal growth temperature 40°C) enzymes of the cellulolytic complex can have temperature optima differing up to 20°C as compared with mesophilic fungi. The temperature optimum of β-glucosidase of A. terreus corresponds to that of the cellulolytic complex of obligatory thermophiles, e.g. Clostridium thermocellum.

From the point of view of enzyme application the thermostability is a very important enzyme characteristic. Thermostable enzymes can be used for saccharification at high temperatures, when high reaction velocities are

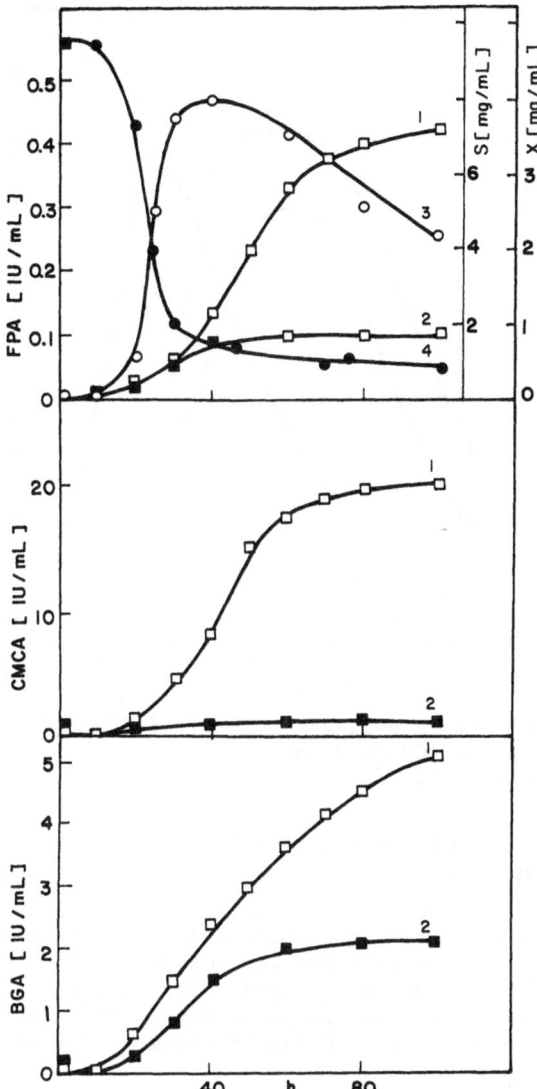

Fig. 1. Cultivation of A. terreus ATG-5 in mineral medium supplemented
with 1% cellulose. a) growth of cells and production of extracellular and bound activity of FP-cellulase (FPA); b) production
of extracellular and bound activity of endoglucanase (CMCA); c)
production of extracellular and bound activity of aryl-β-
glucosidase (βGA). (1) extracellular enzymes; (2) bound enzymes;
(3) dry weight of cells; (4) concentration of cellulose.

reached and the bacterial contamination of the substrates and products of
the enzyme reaction is prevented. Thermostability of β-glucosidase of A.
terreus ATG-5 is illustrated in Figure 3. At 50 and 55°C (pH 4.8) the
enzyme is stable for many hours, and similarly at 60°C, when only a 12%
decrease of the enzyme activity was detected at the beginning of incubation. The half-life of the enzyme assayed at 70°C is 28 min, when the
fastest decrease of the activity is detected. However, after reaching 1/3
of the initial level, β-glucosidase activity remains almost unchanged at
70°C for further 60 min. According to literature data (Workman and Day,

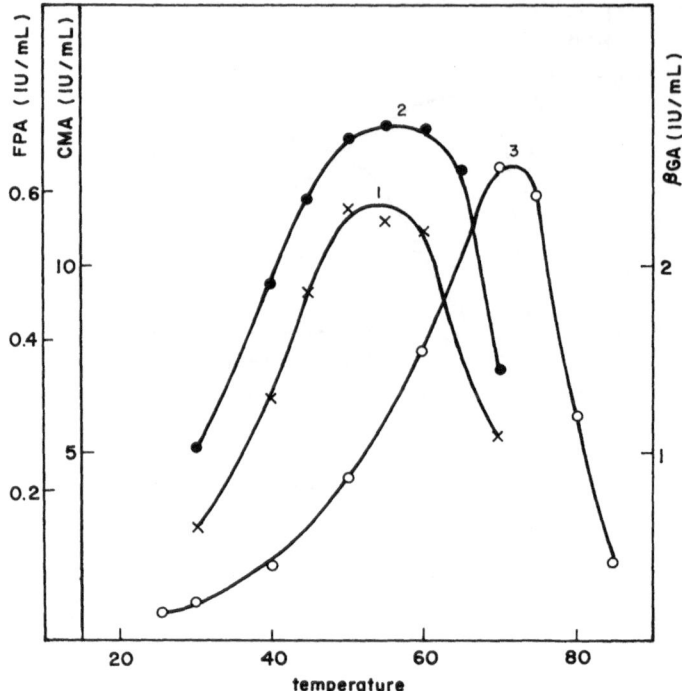

Fig. 2. Effect of temperature on activity of FP-cellulase (FPA; 1),
endoglucanase (CMCA; 2) and aryl-β-glucosidase (βGA; 3) of
A. terreus ATG5.

1982) aryl-β-glucosidase of the strain ATG-5 belongs to the most thermostable β-glucosidases in fungi.

Changed kinetic parameters of the enzyme may be one of the reasons
leading to the increased aryl-β-glucosidase in the mutant ATG-5. When
comparing K_m and V_{max} in the wild and the mutant strain at 50°C (Table 1)
it was shown that both parameters changed. The enzyme affinity to the substrate increased roughly 2.5-fold in the mutant strain, however, V_{max}
simultaneously decreased. Thus, it can be assumed that qualitative changes

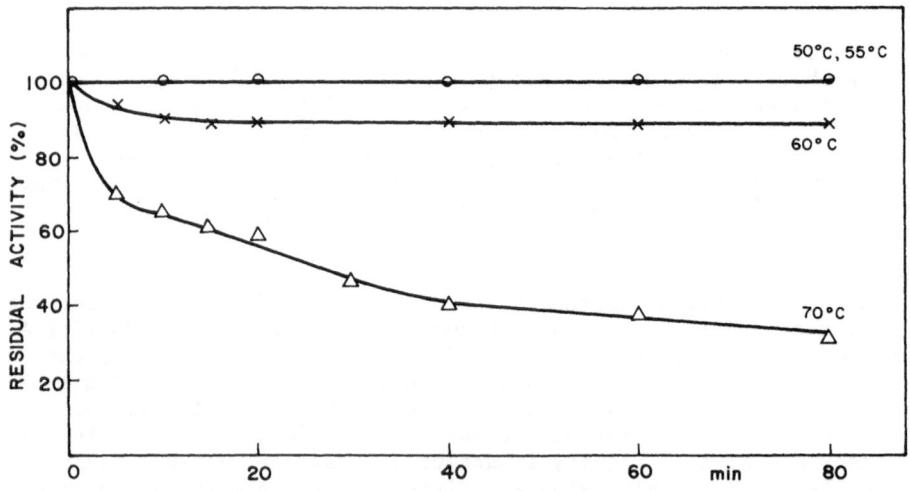

Fig. 3. Aryl-β-glucosidase thermostability of A. terreus ATG-5.

176

Table 1. Physico-chemical Properties of the Extracellular Aryl-β-glucosidase of Wild-type and Mutant Strain of A. terreus.

	AT-0 50°C	ATG-5 50°C	ATG-5 70°C
K_M (mM)	0.59	0.25	0.26
V_{max}	2.00	0.32	1.58
K_i (nM)	2.50*		17.25*
optimum temperature (°C)	70.00		70.00
optimum pH	4.7 - 4.8		4.7 - 4.8

* Determined at 70°C.

of the extracellular aryl-β-glucosidase occurred in the strain ATG-5 and
that enzyme production also increased. This conclusion is also supported
by the fact that the quantity of extracellular proteins increased three
times during growth of the mutant in the mineral medium.

The inhibition of β-glucosidase activity by glucose is one of the
factors influencing negatively saccharification of cellulose. From the
point of view of economy of the saccharification process, the improvement
of cellulolytic microorganisms aimed at the production of β-glucosidase
resistant to the final product appears to be promising. It follows from
literature data that aryl-β-glucosidase of A. terreus belongs to the most
resistant aryl-β-glucoidases (K_i is 3.5 mM glucose; Workman and Day, 1982).
The inhibition constant of aryl-β-glucosidase of the strain ATG-5
determined by us (K_i 17.22 mM glucose; Figure 4) is five times higher than
that presented by Workman and Day (1982) and seven times higher than the
value in the wild strain AT-0 (Table 1). In agreement with literature
data, competitive inhibition of the enzyme activity by glucose is involved
here.

The production of the extracellular aryl-β-glucosidase studied under
constant cultivation conditions in fermentor (30°C, pH 5.0) and in the
presence of various concentrations of crystalline cellulose in the medium
showed that at a higher cellulose concentration (3%) the initial velocity
of the enzyme production is higher than at 1% cellulose but that the total
level of the enzyme produced decreases by 36%. As it is known that the
synthesis of the cellulolytic enzyme complex is subject to catabolite
repression by the products of cellulose degradation, we studied the levels
of reducing saccharides in the medium during growth of the fungus. It
follows from Figure 5 that the level of reducing saccharides is lower in
the medium with 1% cellulose that with 3% cellulose. At the cultivation
time, when the aryl-β-glucosidase production is interrupted, the level of
reducing saccharides reached 0.18-0.20 mg/mL. This concentration has never
been reached in the medium with 1% cellulose. The observed negative effect
of the higher cellulose concentration (3%) on the production of β-glucosidase can be explained by repression of β-glucosidase synthesis by the
accumulating reducing saccharides, glucose in the first place.

The cultivation with feeding of small amounts of cellulose is one of
the possibilities how to decrease the accumulation of the products of
cellulose degradation.

When cultivating the strain ATG-5 with feeding of a small amount of
crystalline cellulose in the medium, the repressive concentrations of
reducing saccharides in the medium were reached 30 h later than in the
cultivation to which the same amount of cellulose was added as a single

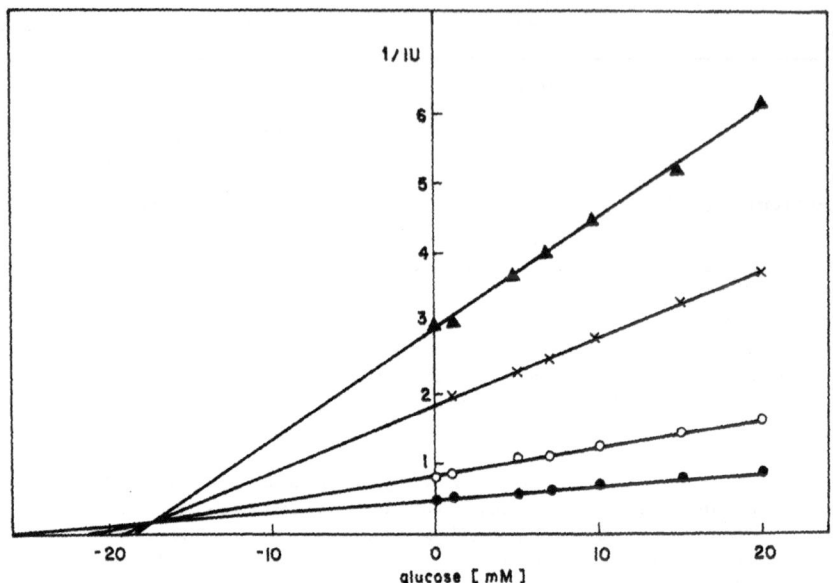

Fig. 4. Dixon plot carried out on assays run in the presence of different
concentrations of glucose and PNPG.

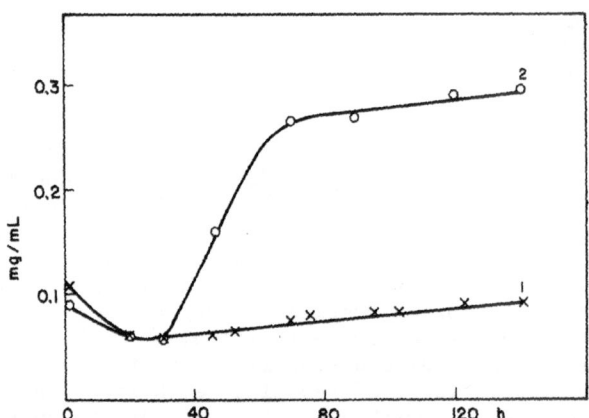

Fig. 5. Reducing saccharides levels in medium during the growth of A.
terreus ATG-5 in a mineral medium supplemented with 1% (1) and
3% (2) cellulose.

dose (Figure 6, curves 2 and 1). As a result, the final activity of aryl-
β-glucosidase in the medium, with feeding of cellulose, increased by up
to 100%.

As according to literature data nitrogen limitation at higher carbon
substrate concentrations may lead to a decreased production of cellulases
due to cessation of the cell growth, we studied the effect of feeding of
cellulose under conditions of feeding of ammonium ions (double fed batch).
The accumulation of reducing saccharides in the medium decreased (Figure 6,
curve 3) and, in addition, the level of extracellular β-glucosidase
increased by additional 53% (Figure 7, curves 2 and 1). As specific

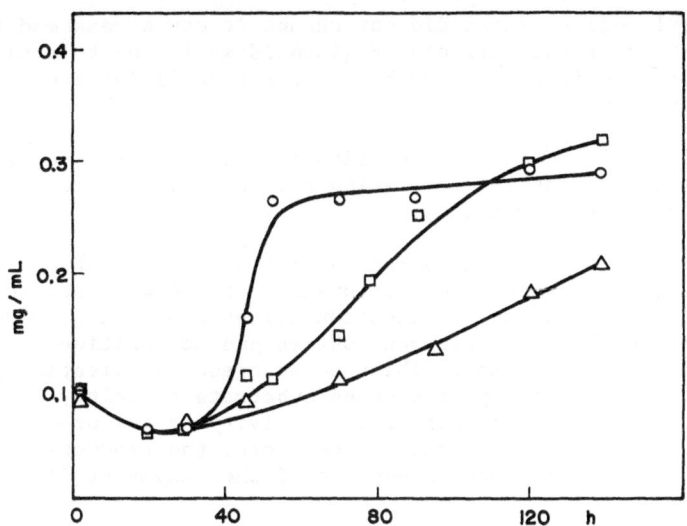

Fig. 6. Effect of fed batch (source of carbon) and double fed batch
(source of carbon and source of nitrogen) on the concentration of
reducing saccharides in medium during the growth of A. terreus
ATG-5. (1, ○) 3% cellulose (added as one dose); (2, □) 3%
cellulose fed batch; (3, △) 3% cellulose double fed batch.

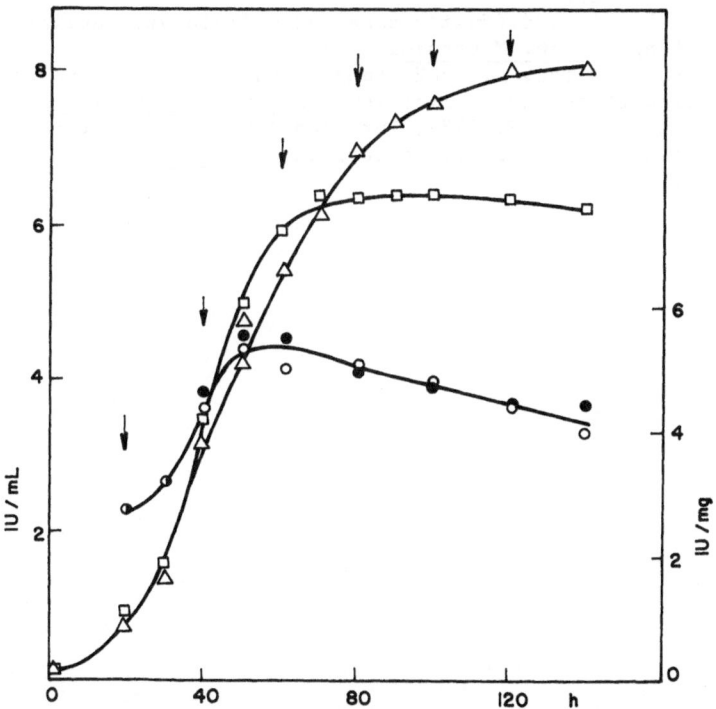

Fig. 7. Effect of fed batch (source of carbon) and double fed batch
(source of carbon and source of nitrogen) cultivation on the
production of aryl-β-glucosidase in A. terreus ATG-5. (1, □)
extracellular enzyme activity in fed batch; (2, △) extracellular
enzyme activity in double fed batch; (3, ○) specific enzyme
activity referred to medium protein in fed batch; (4, ●) specific
enzyme activity in double fed batch.

activities of aryl-β-glucosidase did not change it may be assumed that the increased production of extracellular β-glucosidase is due to a higher biomass production and increased synthesis of extracellular proteins (Figure 7, curves 3 and 4).

The maximal levels of aryl-β-glucosidase, i.e. 21 IU/mL, were reached in complex media containing meal, to which crystalline cellulose was gradually added during the cultivation.

In conclusion it may be stated that a high-production strain A. terreus ATG-5 was obtained by means of improvement of the wild strain of A. terreus and optimization of cultivation conditions consisting primarily in the way of feeding carbon and nitrogen sources and composition of the cultivation medium. The strain exhibits advantageous application properties, i.e. high thermostability, increased substrate affinity, seven times lower sensitivity to the inhibition of the activity by the product of the reaction and a high enzyme production. Therefore, the production strain A. terreus ATG-5 belongs to the best producers of the enzyme studied.

REFERENCES

Araujo, A., and D'Souza, J. 1980, Production of biomass from enzymatic hydrolysate of agricultural waste, J. Ferment. Technol., 58:399.

D'Souza, J., and Volfová, O. 1982, The effect of pH on the production of cellulases in Aspergillus terreus, Eur. J. Appl. Microbiol. Biotechnol. 16:123.

Tansey, M. R. 1971, Agar-diffusion assay of cellulolytic ability of thermophilic fungi, Arch. Microbiol., 77:1.

Woodward, J., and Wiseman, A. 1982, Fungal and other β-D-glucosidases - their properties and applications, Enzyme Microb. Technol., 4:73.

Workman, W.E., and Day, D.F. 1982, Purification and properties of β-glucosidase from Aspergillus terreus, Appl. Environ. Microbiol., 44:1289.

LOCALIZATION OF β-D-GLUCOSIDASE ACTIVITY IN

STREPTOMYCES GRANATICOLOR ETH 7347

M. Jirešová, Z. Dobrová, J. Janeček and J. Náprstek

Department of General Microbiology
Institute of Microbiology
Czechoslovak Acad. Sci., Prague 4

The fact that streptomycetes represent a higher degree of all organization among bacteria makes them an interesting model for studying regulatory systems as compared with other prokaryotes where the regulation of protein synthesis has been studied in more detail. Inducible enzyme synthesis is one of the best-known types of regulation of protein synthesis. In streptomycetes most attention was paid to the inducible synthesis of β-D-galactosidase (EC 3.2.1.23) (Dan and Szabó 1973, Chaterjee and Vining 1982a). Chaterjee and Vining (1981, 1982b) studied also the induction of α- and β-D-glucosidase (EC 3.2.1.20,21) in Streptomyces venezuelae. Results obtained in the study of the induction of β-D-glucosidase in S. granaticolor are presented in this communication. This enzyme is well suited for the study of protein synthesis regulation because it is very easy to assay and exhibits a high degree of induction. Moreover, β-D-glucosidase is important in the bioconversion of cellulose to glucose (Ishague and Kluepfeld 1980). The enzyme not only produces glucose but also alleviates the cellobiose inhibition allowing, the cellulolytic enzymes to function more efficiently.

The microorganism was cultivate in 500 mL flasks containing maximally 70 mL of the medium. A minimal cultivation medium M56 (Cohn and Monod 1951) containing 1% Casamino acids was supplemented with 1% glycerol. Cultivation was done on a reciprocal shaker at 27°. Synthesis of β-D-glucosidase was induced by adding the following inducers: cellobiose (4-0-β-D-glucopyranosyl-D-glucose), salicin (saligenin-β-D-glucopyranoside), arbutin (p-hydroxyphenyl-β-D-glucopyranoside), methyl β-D-glucopyranoside (MeG) carboxymethyl-cellulose (CMC) and microcrystalline cellulose (MC). The culture was centrifuged for 15 min at 15000 g in the cold. The sediment was washed and resuspended in 10 mM Na-phosphate buffer (pH 6,4). The cells were disintegrated by vortexing with an equal volume of wetted ballotini beads. The β-D-glucosidase activity was routinely assayed with p-nitrophenyl-β-D-glucopyranoside (PNPG) as substrate. The absorbance was read at 425 nm. The β-D-glucosidase activity was determined in the medium, in washed intact cells and in cell-free extract - (Figure 1). β-D-Glucosidase activity in the cell-free extract and in the cultivation medium was partially purified before electrophoresis with ammonium sulphate (AmS). The protein fraction sedimenting at 60-70% saturation contained the majority of the enzyme activity.

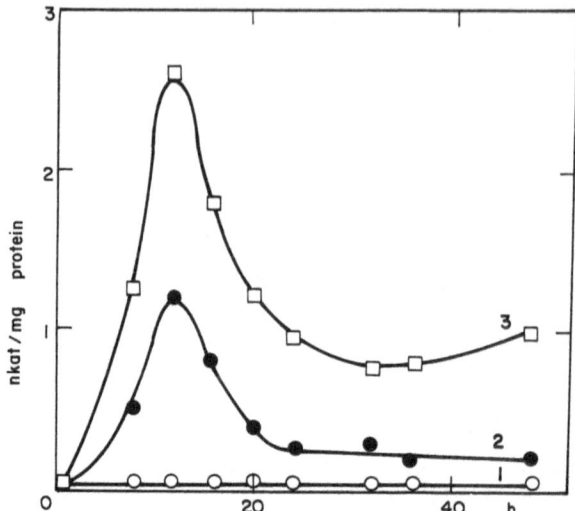

Fig. 1. Localization of β-D-glucosidase. The enzyme synthesis was induced by adding 5 mM methyl-β-D-glucopyranoside to the culture growing in the medium containing glycerol. At time intervals a part of the culture was centrifuged and enzyme activity was assayed in the medium (1), in washed intact cells (2) and in the cell-free extract prepared from the same cells (3) homogenized with ballotini.

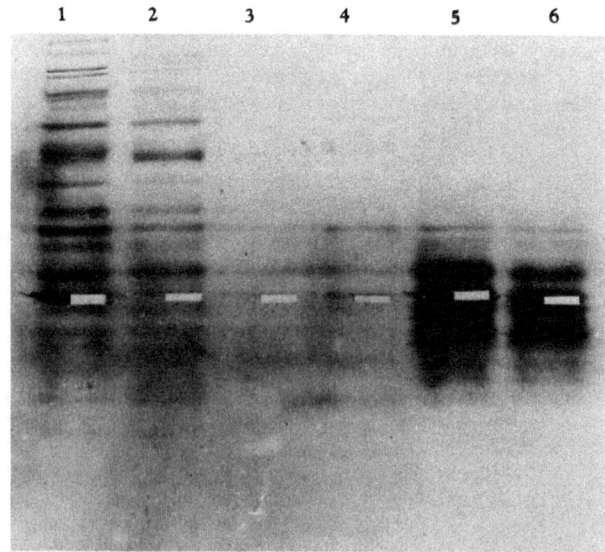

Fig. 2. Gel electrophoresis (7,5% acrylamide) of cell-free extracts of Streptomyces granaticolor after induction with different inducers: arbutin (lane 1), methyl-β-D-glucosidase (lane 2), cellobiose (lane 3), salicin (lane 4), cellobiose + glycerol (lanes 5,6) – the protein band which exhibit the β-D-glucosidase activity was partially purified by elution from preparative polyacrylamide gel. White bands at the right side of each lane indicate the position of β-D-glucosidase activity.

Acrylamide gel electrophoresis was carried out according to Maurer (1971) with slight modifications. To visualize the β-D-glucosidase activity unfixed gels were immersed in 10 mM PNPG (in Na-phosphate buffer pH 6,4; at room temperature) for 30 min. The formation of a yellow band indicated the liberation of p-nitrophenol and thus the location of β-D-glucosidase. Then the gels were stained with 0.25% Coomassie Brilliant Blue to visualize different protein bands (Figure 2). When CMC of MC (and C as a control) served as inducers the enzyme activities were prepared in a different way:

a) The fermentation medium devoid of cells was brought to 70% saturation with AmS to precipitate the protein. The suspension was centrifuged (20000 g 30 min) and the pellet dissolved in a minimum amount of Na-phosphate buffer (pH 6,4 - Figure 3).

b) The pelleted cells were broken by freezing-thawing in the presence of lysozyme. DNase was subsequently added. The suspension was centrifuged at 20000 g for 40 min. The treatment liberated the enzyme activity into the supernatant (Figure 4).

c) Some of the enzyme activity was also found in the pellet after the previous treatments. The pellet consists not only of the cell debris but also of CMC or MC (Figure 5).

In all these fractions the PNPGase activity was determined (Table 1) and thereafter analyzed in 7,5% polyacrylamide gels (Figures 3,4,5).

When Arbutin, MeG, Cellobiose and Salicin served as inducers, the β-D-glucosidase activity (PNPG used as substrate) was localized in the cytoplasm and in the periplasm. With CMC and MC as inducers the enzyme activity was found mostly in medium. The enzyme(s) formed after induction with different inducers possess the same electrophoretic mobility in the 7,5% polyacrylamide gel. The possibility that the enzymic activities are

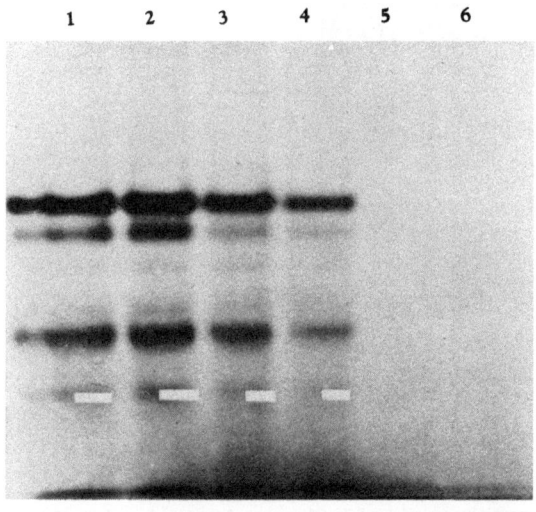

Fig. 3. Gel extrophoresis (7,5% acrylamide of cultivation medium of Streptomyces granaticolor after induction with CMC (lane 1,2) MC (lane 3,4), and cellobiose (5,6). White bands at the right side of each lane indicate the position of β-D-glucosidase activity.

different due to glycosylation of the protein moiety, or because of different substrate specificity, has to be proved.

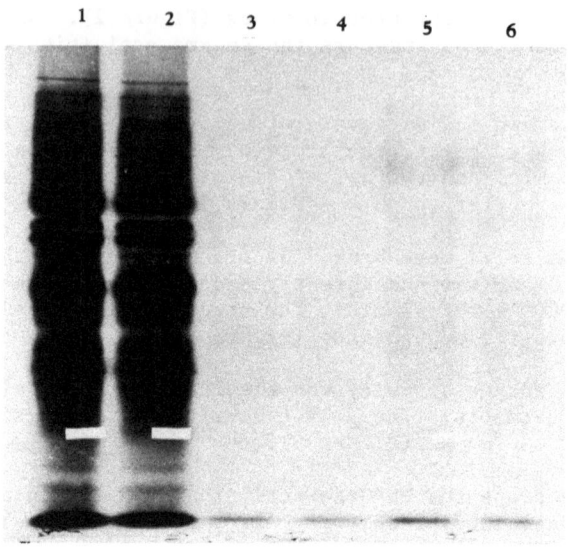

Fig. 4. Gel electrophoresis (7,5% acrylamide) of cell-free extracts of Streptomyces granaticolor after induction with cellobiose (lane 1,2), CMC (lane 3,4) and MC (lane 5,6). White bands at the right side of each lane indicate the position of β-D-glucosidase activity.

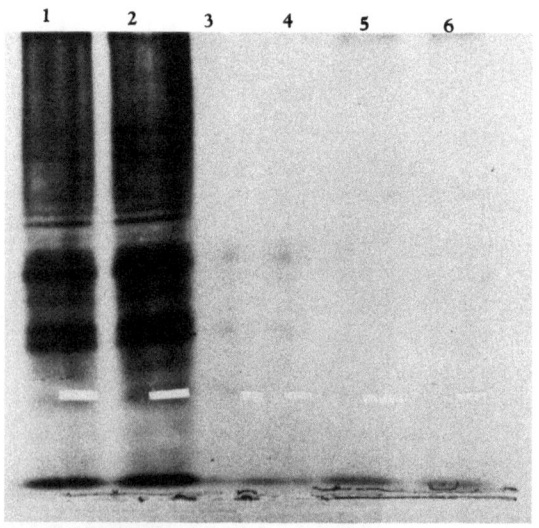

Fig. 5. Gel electrophoresis (7,5% acrylamide) of cell debris + CMC or MC of Streptomyces granaticolor after induction with cellobiose (lane 1,2) CMC (lane 3,4) or MC (lane 5,6). White bands at the right side of each lane indicate the position of β-D-glucosidase activity.

Table 1. The Effect of Different Inducers (Cellobiose, CMC and MC) on the Synthesis and Localization of β-D-glucosidase Activity.

Activity (n kat/mg protein) determined in	Enzyme activity in the presence of		
	cellobiose	CMC	MC
medium	0	0,113	0,115
cell debris	0,654	0,043	0,047
cell-free extract	1,298	0	0,14

REFERENCES

Chatterjee, S., Vining, L. C., 1981, Nutrient utilization in actinomycetes. Induction of β-glucosidase in Streptomyces venezuelae, Can. J. Microbiol. 27:639.
Chatterjee, S., Vining, L. C., 1982a, Catabolite repression in Streptomyces venezuelae, Induction of β-galactosidase, chloramphenicol production, and intracellular cyclic adenosine 3',5' monophosphatase concentrations, Can. J. Microbiol. 28:311.
Chatterjee, S., Vining, L. C., 1982b, Glucose suppression of β-glucosidase activity in chloramphenicol - producing strain of Streptomyces venezuelae, Can. J. Microbiol. 28:593.
Cohn, M., Monod, J., 1951, Purification et properties de la β-galactosidase (lactase) d´Escherichia coli, Biochim. Biophys. Acta 7:1953.
Dán, A., Szabo, G., 1973, Induced production of β-galactosidase in Streptomyces griseus, Acta Biol. Acad. Sci. Hung. 24:1.
Ishague, M., Kluepfeld, D., 1980, Cellulase complex of a mesophilic Streptomyces strain, Can. J. Microbiol. 26:183.
Maurer, H. R., 1971, Disc electrophoresis and related techniques of polyacrylamide gel electrophoresis, Walter de Gruyter and co., Berlin, New York.

DIFFERENTIATION OF GLYCANASES OF MICROBIAL CELLULOLYTIC
SYSTEMS USING CHROMOGENIC AND FLUOROGENIC SUBSTRATES

Peter Biely

Institute of Chemistry, Center for Chemical Research
Slovak Academy of Sciences, 842 38 Bratislava
Czechoslovakia

INTRODUCTION

Microbial cellulolytic systems are composed of several enzyme components attacking cellulose, cello-oligosaccharides and xylan. An understanding of the nature of various enzymes and their substrate specificities
is important in developing processes that have industrial potential.
Reliable enzyme differentiation is needed especially if genes for some
enzymes are to be cloned. This need has been recently partially filled by
introduction of sensitive methods for detection and identification of
endo-1,4-β-glucanases and endo-1,4-β-xylanases employing the binding of
Congo Red to carboxymethylcellulose and xylan [1-3], or employing the
soluble covalently dyed hydroxyethylcellulose [4] and xylan [4,5].
Additional progress in the area of differentiation of cellulolytic enzymes
has been achieved by the use of several fluorogenic substrates, 4-methyl-
umbelliferyl β-glycosides of glucose, cellobiose [6] and lactose [6,7]. In
this contribution we would like to report that a suitable combination of
chromogenic and fluorogenic substrates for the detection of enzyme activities in separation gels (after electrophoresis or isoelectric focusing) or
in solid growth selection media, permits a rapid and efficient differentiation of glycanases and some glycosidases according to their substrate
specificities.

SUBSTRATES

Chromogenic substrates: Ostazin Brilliant Red hydroxyethylcellulose
(OBR-HEC), a soluble covalently dyed cellulose derivative, a substrate
specific for endo-1,4-β-glucanases (cellulases) [4]; Remazol Brilliant Blue
xylan (RBB-X), a soluble covalently dyed beechwood 4-O-methyl-D-glucurono-
D-xylan, a specific substrate for endo-1,4-β-xylanases [4]. Both substrates are available on request from Chemapol Ltd., Praha, Czechoslovakia.

Fluorogenic substrates: 4-Methylumbelliferyl β-D-glucoside (Umb-G), a
substrate for cellobiase and aryl β-glucosidase [6]; 4-methylumbelliferyl
β-cellobioside (Umb-G$_2$), a substrate for cellobiohydrolase (exo-1,4-β-
glucanase), cellobiase and certain types of endo-1,4-β-glucanases [6];
4-methylumbelliferyl β-lactoside (Umb-L), a substrate for cellobiohydrolase
and certain types of endo-1,4-β-glucanases [6,7]. The glycosides are
commercially available (e.g., from Sigma) and can also be easily synthesized [8].

PRINCIPLES OF DETECTION

The detection of endoglycanase activities by chromogenic polymeric substrates is based on changes in the diffusion rates and in solubility in the presence of organic solvents of the original substrate due to hydrolysis (Figure 1). Selective removal of depolymerized dyed substrates from transparent agar replicas, resulting in a destaining of areas of enzyme localization, is achieved by solvents which neither solubilize nor precipitate the original non-degraded polysaccharide [4]. In solid growth media, supplied with covalently dyed polysaccharides the production of the corresponding endoglycanase is seen as an appearance of a decolored zone around the point of inoculation or around the cell colonies [9].

The basis for the detection of enzyme activities using 4-methylumbelliferyl glycosides is the release of the glycon, 4-methylumbelliferone, which shows an intense fluorescence in UV light ($\sim$ 350 nm) [6]. When a glycoside is present in a solid selection medium fluorescent zones appear under UV light around points of inoculation or around colonies of cells secreting the corresponding glycosidase. Umb-L as an example of a fluorogenic substrate, and products of its hydrolysis are shown in Figure 2.

MODE OF APPLICATION OF SUBSTRATES

The detection of enzymes in separation gels (flat-bed gels are most suitable) is conveniently done by the overlay technique using thin-layer (0.75-1.0 mm) 2% agar gels containing 0.5% (w/w) of a chromogenic substrate [4] or 1.0-5.0 mM 4-methylumbelliferyl glycoside. The pH of the agar detection gel should be selected with respect to the pH of the separation gel, so that a value close to the pH optimum of the followed enzyme is

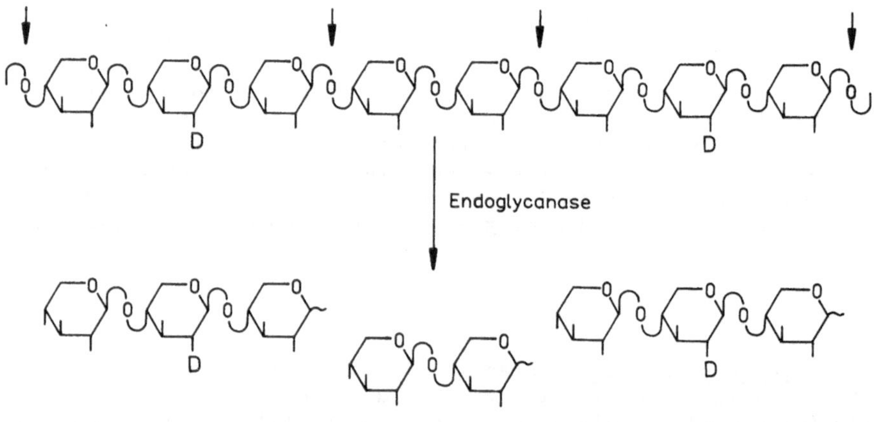

Endoglycanase

Fig. 1. Schematic representation of a 1,4-β-linked polysaccharide with covalently attached dye (D) and the products of its degradation by an endo-1,4-β-glycanase. The non-hydrolyzed polymer is water-soluble, precipitable by organic solvents from aqueous media, but non-precipitable under the same conditions when present in 2% agar gel. The low molecular weight fragments resulting from enzymic cleavage (arrows) remain soluble in the presence of organic solvents, so that they can be selectively washed out from the agar gel. Due to an increased rate of diffusion of the liberated fragments in comparison to the starting polymer, the enzyme action in a substrate-containing gel is accompanied by changes in the intensity and shade of the color background.

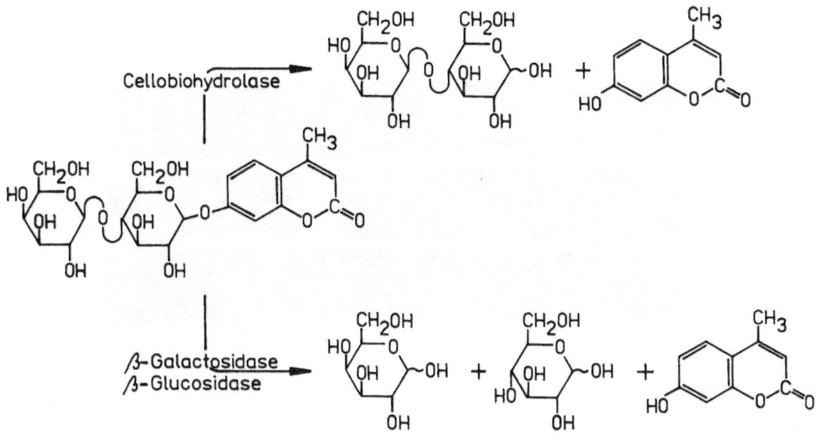

Fig. 2. 4-Methylumbelliferyl β-D-lactoside and its hydrolysis to D-lactose and 4-methylumbelliferone by cellobiohydrolase (or by some endo-1,4-β-glucanases), and to D-galactose and D-glucose in two steps by β-galactosidase and β-glucosidase.

obtained after the separation and the detection gels are brought into contact. Two enzyme activities can be detected simultaneously when two substrates, one chromogenic and one fluorogenic, are present in the gel. Flooding of the separation gels with a solution of a fluorogenic glycoside was employed by other authors [6].

Solid selection growth media are prepared with 0.2% (w/w) OBR-HEC or RBB-X [9]. The substrates can be sterilized by autoclaving either separately in the dry state [9] or dissolved in complete media. Fluorogenic glycosides used at 0.5-2.5 mM concentration can be sterilized by autoclaving too; however, their stability should be checked. Sterilization by filtration is recommended when more acidic media are used. Use of two substrates, a covalently dyed polysaccharide in a combination with a fluorogenic glycoside, in solid selection media greatly accelerates the screening for cells which express the genes of cellulolytic enzymes. Selection media containing more than one fluorogenic glycoside can also be designed. In such a case an additional cell differentiation step is required but the saving of the number of the selection plates is enormous.

ENZYME DIFFERENTIATION

OBR-HEC is attacked only by endo-1,4-β-glucanases. Some endo-1,4-β-glucanases are not specific for cellulose and hydrolyze also RBB-X. Besides different affinity towards xylan, endo-1,4-β-glucanases can be differentiated by their ability to hydrolyze Umb-L and Umb-G_2, releasing the disaccharides and the aglycon [6,7]. However, Umb-L and Umb-G_2 are cleaved in the same pattern by cellobiohydrolase (exo-1,4-β-glucanase liberating cellobiose) [6,7,10]. Umb-L is a more convenient substrate for the detection of cellobiohydrolase than Umb-G_2 because the latter also serves as a substrate for β-glucosidase (cellobiase). On the other hand, the fluorescent aglycon of Umb-L can also be liberated in two steps by β-galactosidase and β-glucosidase (Figure 2). Therefore, the use of Umb-L, particularly for the selection purposes, is conditioned by the absence of one of the glycosidases.

Figure 3 shows an example of enzyme analysis of the cellulolytic system of the fungus Schizophyllum commune which has been extensively

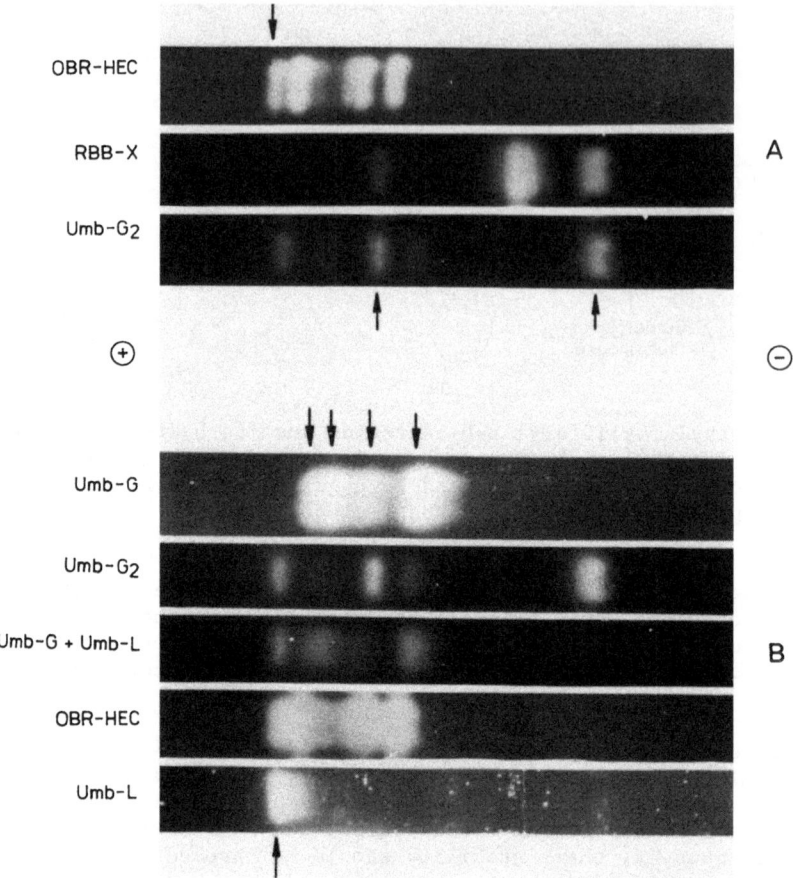

Fig. 3. Cellulolytic system of <u>Schizophyllum commune</u> resolved by thin-layer electrophoresis in agarose gel (pH 8.6). Enzymes were detected using agar replicas containing the chromogenic and fluorogenic substrates as indicated (for abbreviations see the text). Part A: a downward arrow marks one of the six endo-1,4-β-glucanase components that attacks Umb-G_2; upward arrows point to endo-1,4-β-xylanase components which hydrolyze Umb-G_2 without attacking OBR-HEC. Part B: downward arrows point to four components of β-glucosidase which liberate the fluorescent aglycon from both Umb-G and Umb-G_2; the upward arrow marks the endo-1,4-β-glucanase component hydrolyzing Umb-L.

studied by Canadian workers [11]. The proteins secreted by the fungus during growth on cellulose were separated by thin-layer electrophoresis in agarose gel and the enzyme activities were localized in the gel by the overlay technique using five different substrates. Some of the resolved enzyme components hydrolyzed only one substrate, others hydrolyzed two or three substrates. One of the six detected endo-1,4-β-glucanase components attacked Umb-G_2 and Umb-L, indicating the presence of at least two different types of glucanases. Unexpected was the coincidence of two endo-1,4-β-xylanase components with the activity hydrolyzing Umb-G_2 (Umb-L was also hydrolyzed but at a much lower rate) (Figure 3A). The finding is not original, because xylanases capable of hydrolyzing β-cellobiosides have been recently identified among the cellulolytic enzymes of <u>Dichomitus</u>

squalens [12]. Of four β-glucosidase components detected with Umb-G, at least three hydrolyzed Umb-G_2 but none of them attacked Umb-L (Figure 3B).

The main difference in the enzyme composition of S. commune in comparison to that of Trichoderma reesei [1,10] seems to be the absence of non-specific endo-1,4-β-glucanases hydrolyzing also RBB-X and Umb-L. All endo-1,4-β-glucanases and endo-1,4-β-xylanases of S. commune appear to be more specific for the corresponding polysaccharide. Results of a two-dimensional resolution of the cellulolytic enzymes of S. commune will be published elsewhere.

The above example of enzyme analysis together with the results obtained from studies of other systems, point to a great potential of using covalently dyed polysaccharides in combination with fluorogenic glycosides for differentiation of cellulolytic enzymes. This fact is illustrated in Table 1.

Table 1. Differentiation of Cellulolytic Enzymes According to their Substrate Specificities using Covalently Dyed Hydroxyethyl-cellulose (OBR-HEC) and xylan (RBB-X) and 4-methylumbelliferyl β-glycosides as Substrates. For abbreviations see the text.

Enzyme	Substrate hydrolyzed
Endo-1,4-β-glucanase (specific, non-hydrolyzing xylan)	OBR-HEC
Endo-1,4-β-glucanase (specific, non-hydrolyzing xylan, but hydrolyzing β-cellobiosides and β-lactosides)	OBR-HEC, [a]Umb-G_2, [a]Umb-L
Endo-1,4-β-glucanase (non-specific, hydrolyzing xylan and β-cellobiosides and β-lactosides)	OBR-HEC, RBB-X, [a]Umb-G_2, [a]Umb-L
Endo-1,4-β-xylanase (specific, non-hydrolyzing cellulose)	RBB-X
Endo-1,4-β-xylanase (specific, non-hydrolyzing cellulose but hydrolyzing β-cellobiosides)	RBB-X, [a]Umb-G_2 ([a]Umb-L)
Endo-1,4-β-xylanase (non-specific)	RBB-X, OBR-HEC
Cellobiohydrolase (exo-1,4-β-glucanase)	[a]Umb-L and/or [a]Umb-G_2
Cellobiase (β-glucosidase)	[b]Umb-G_2, Umb-G
Aryl β-glucosidase (non-hydrolyzing cellobiose)	Umb-G

[a] Hydrolysis at the second glycosidic bond from the non-reducing end.
[b] Hydrolysis at both glycosidic bonds.

REFERENCES

1. R. M. Theater, and P. J. Wood, Use of Congo Red-Polysaccharide Inter-actions in Enumeration and Characterization of Cellulolytic Bacteria from the Bovine Rumen, App. Environ. Microbiol., 43:777 (1982).

2. P. Béguin, Detection of Cellulase Activity in Polyacrylamide Gels using Congo Red Stained Agar Replicas, Anal. Biochem., 131:333 (1983).

3. C. R. MacKenzie, and R. E. Williams, Detection of Cellulase and Xylanase Activity in Isolelectric Focusing Gels using Plastic-Film Supported Agar Substrate Gels, Can. J. Microbiol., 30:1522 (1984).

4. P. Biely, O. Markovič, and D. Mislovičová, Sensitive Detection of Endo-1,4-β-glucanases and Endo-1,4-β-xylanases in Gels, Anal. Biochem., 144:147 (1985).

5. Y. Bertheau, E. Madgidi-Hervan, A. Kotoujansky, C. Nguyen-The, T. Andro, and A. Coleno, Detection of Depolymerase Isoenzymes after Electrophoresis, Electrofocusing, or in Titration Curves, Anal. Biochem., 139:383 (1984).

6. H. van Tilbeurgh, and M. Claeyssens, Detection and Differentiation of Cellulase Components using Low Molecular Mass Fluorogenic Substrates, FEBS Letters, 187:283 (1985).

7. P. Biely, and O. Markovič, Resolution of Glycanases of Trichoderma reesei with Respect to Cellulose and Xylan Degradation, Bio/-Technology, in press.

8. N. Constantzas, and J. Kocourek, Glycoside des 4-Metylumbelliferones, Coll. Czechoslov. Chem. Commun., 24:1099 (1959).

9. V. Farkaš, M. Lišková, and P. Biely, Novel Media for Detection of Microbial Producers of Cellulase and Xylanase, FEMS Microbiol. Letters, 28:137 (1985).

10. M. Hrmová, P. Biely, and M. Vršanská, Specificity of Cellulase and β-xylanase Induction in Trichoderma reesei QM 9414, Arch. Microbiol., 144:307 (1986).

11. G. E. Willick, R. Morosoli, V. L. Seligy, M. Yaguchi, and M. Desrochers, Extracellular Proteins Secreted by the Basidiomycete Schizophyllum commune in Response to Carbon Source, J. Bacteriol., 159:294 (1984).

12. X. Rouau, and E. Odier, Purification and Properties of Two Enzymes from Dichomitus squalens which Exhibit Both Cellobiohydrolase and Xylanase Activity, Carbohydr. Res., 145:279 (1986).

PART V
OTHER ENZYMES

BIOSYNTHESIS AND PROPERTIES OF EXTRACELLULAR PULLULANASE

FROM BACILLUS STEAROTHERMOPHILUS G-82

M. S. Kambourova and E. I. Emanuilova

Institute of Microbiology
BAS, 1113 Sofia, Bulgaria

INTRODUCTION

The production of thermostable enzymes, catalyzing reactions at high temperatures, is one of the most attractive features of thermophilic microorganisms. In order to select producers of thermostable hydrolases, different aerobic thermophilic bacterial strains were isolated from water, soil and organic material samples collected from Bulgarian hot springs environment. Some of the properties of the isolated strains were a subject of our previous work [1]. The aim of the present paper was to characterize and identify the strain producing thermostable pullulanase as well as to establish the optimum conditions for enzyme production and to study some of the enzyme's properties.

MATERIALS AND METHODS

Media and cultivation

The isolated thermophilic strain was cultivated in a liquid medium N1, containing (in %): 0.5 peptone; 0.5 yeast extract; 20 potato extract [2]; pH 7.8-8.0. For the maintenance and preservation of the strain, the same nutrient medium, containing 2,5% agar was used. In order to study the influence of different carbohydrates on pullulanase production, the strain was grown in medium N1A, containing (in %): 0.1 carbohydrate; 0.5 peptone; 0.2 yeast extract; pH 7.8-8.0. Cultivation was carried out in 100ml flasks with 14 ml of medium on a New Brunswick G-76 shaker at 55°C. The growth of the cultures and pullulanase activity were determined after 12 h of cultivation. The investigations in 750 ml jar fermentor (Bioflo, New Brunswick Sci. Co. Inc.) were performed in medium N2, containing (in %): 0.1 starch; 0.3 peptone; 0.3 yeast extract; pH 7.8-8.0.

Bacterial Cell Concentration

It was determined by reading optical density (OD) of cultures at 660 nm. It was established that a unit of optical density corresponded to 0.906 mg of dry cells per ml.

Assay of Pullulanase

The method of Suzuki and Chishiro [3] was used.

Enzyme Purification

The culture filtrate was concentrated using the Reichelt ultra-filtration system. The ultrafiltrate was saturated with 80% of $(NH_4)_2SO_4$. Gel chromatography was carried out on Sephadex G-75, after which DEAE cellulose chromatography was followed by Sephadex G-100 re-chromatography. The electrophoretic method of Laemmli [4] and staining method of Vesterberg et al. [5] were used. Molecular weight was estimated on Sephadex G-100, using standard markers.

RESULTS AND DISCUSSIONS

Main Characteristics of the Strain

The cells were Grampositive, motile, rod shaped (0.5-0.8 by 1.6-4.8 μm). The spores were found to be oval and terminally located. Colonies formed on medium N1 were round, with smooth surface, with convex center, beige in color, past-like. The strain grew at temperatures ranging from 32° or 68°, exhibiting optimum at 60°C. The pH range for growth was from 6.2 to 8.2 with an optimum at pH 7.5. The strain was a facultative anaerobe, it reduced nitrates, hydrolyzed gelatin and casein, did not form catalase, had ammonia positive and methyl-red negative reactions and did not haemolyze blood agar. The strain studied formed acid without gas while assimilating the following carbohydrates: sucrose, glucose, xylose, maltose, raffinose, galactose, cellobiose, starch and inositol. The bacterium was able to grow on 3% NaCl. Based on its characteristics the isolated thermophile was identified as belonging to the species Bacillus stearothermophilus.

Table 1. Influence of Different Carbohydrates on Biomass and Pullulanase Production of Bacillus stearothermophilus G-82.

Carbohydrate	Biomass mg/ml	Enzyme activity $U.10^{-2}/ml$	Specific enzyme productivity $U.10^{-2}/mg$
Polysaccharides			
Amylopectin	0.770	31.6	41.0
Pullulan	0.734	27.8	36.1
Starch	0.543	18.2	33.5
Dextrin	0.842	14.5	16.7
Glycogen	0.597	4.4	7.4
Trisaccharides			
Raffinose	0.688	1.2	1.7
Disaccharides			
Maltose	0.643	1.7	2.7
Sucrose	0.761	1.1	1.4
Monosaccharides			
Galactose	0.589	1.0	1.7
Fructose	0.616	0.6	0.9
Glucose	0.616	0	0

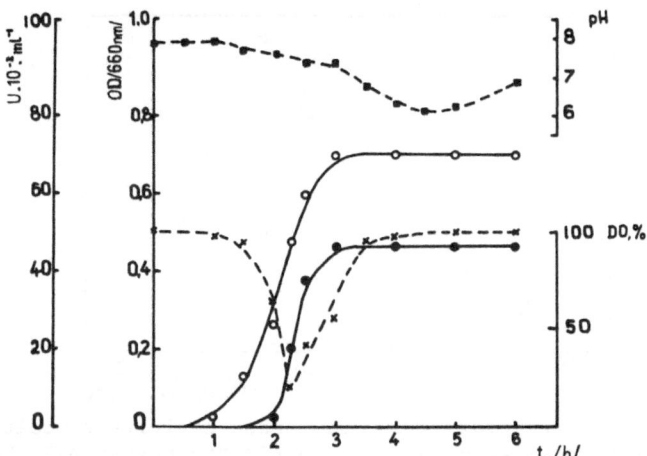

Fig. 1. Pullulanase production (●——●) during growth (o——o) of Bacillus stearothermophilus G-82; pH (■---■); dissolved oxygen % (x---x). Growth conditions: temperature 55°C, aeration 0.5 vvm, agitation 350 min⁻¹, medium N2.

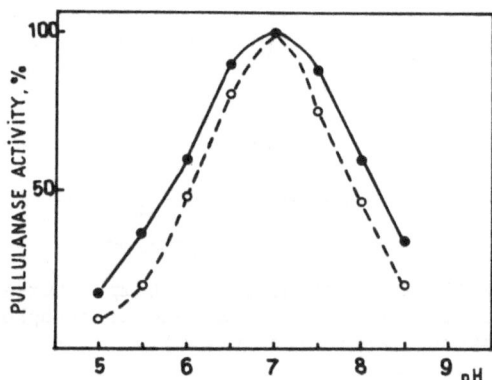

Fig. 2. Effect of pH on pullulanase activity of culture supernatant (●——●) at 65°C and of the electrophoretically homogeneous enzyme preparation (o---o) at 55°C.

Pullulanase Production

It was established that the optimum temperature for pullulanase production by Bacillus stearothermophilus G-82 was 55°C and initial pH of the media used was 8.0-8.2. As shown in Table 1, enzyme synthesis was significantly enhanced by amylopectin, pullulan and starch. When the strain was grown in 350 ml of medium N2, in jar fermentor Bioflo, the maximum enzyme productivity of 0,464U.ml⁻¹ could be achieved after 3 h of cultivation (Figure 1). The pullulanase activity of the culture continued to increase linearly until the cessation of cell growth. These results indicated that enzyme synthesis was growth-associated when medium N2 was used. Compared with B. stearothermophilus KP 1064 isolated by Suzuki and Chishiro [3], the strain B. stearothermophilus G-82 demonstrated higher pullulanase productivity for a shorter cultivation time.

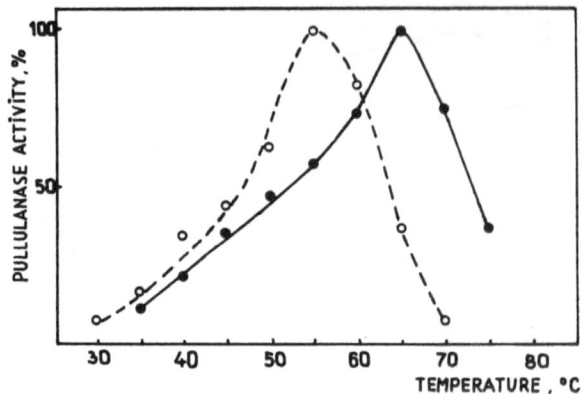

Fig. 3. Effect of temperature on pulluanase activity of culture supernatant (●——●) and of the electrophoretically homogeneous enzyme preparation (○---○).

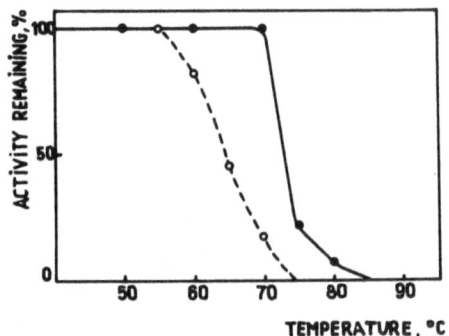

Fig. 4. Stability of pullulanase at different temperatures for 10 min at pH 7.0: culture supernatant (●——●) and electrophoretically homogeneous enzyme preparation (○---○).

Pullulanase Properties

It was established that the purified electrophoretically homogeneous enzyme had a molecular weight of 53 000. Figures 2, 3 and 4 show some of the enzyme's properties in the culture supernatant in comparison with those of the purified enzyme. The optimum of the enzyme activity in the supernatant was at 65°C, pH 7.0 and it was 100% retained for 10 min at 70°, while the purified electrophoretically homogeneous enzyme had an optimum at 55°C, pH 7.0 and it was 100% retained for 10 min up to 55°C.

The results of these experiments show that the strain Bacillus stearothermophilus G-82 is a promising producer of thermostable pullulanase and is of interest for further investigation.

REFERENCES

1. M. S. Kamburova and E. I. Emanuilova, On isolation of Thermophilic Aerobic Bacteria from Hyperthermal Springs in Bulgaria and Studies on Some of their Properties, Acta Microbiol. Bulg., 17:3 (1985).

2. L. G. Loginova and L. A. Egorova, "New forms of Thermophilic Bacteria", Nauka, Moscow (1973).

3. Y. Suzuki and M. Chishiro, Production of Extracellular Thermostable Pullulanase by an Amylolytic Obligated Thermophilic Soil Bacterium Bacillus stearothermophilus KP 1064, European j. Appl. Microbiol. Biotechnol., 17:24 (1983).

4. U. K. Laemmli, Cleavage of Structural Proteins During the Assembly of the Head of Bacteriophage T4, Nature 227:680 (1970).

5. O. Vesterberg, L. Hansen, and A. Sjosten, Staining of Proteins after Isoelectric Focusing in Gels by New Procedures, Biochim. Biophys. Acta, 491:160 (1977).

CONTINUOUS PRODUCTION OF THE EXTRACELLULAR HYDROLYTIC SYSTEM

BY IMMOBILIZED MYCELIA OF <u>ALTERNARIA TENUISSIMA</u>

V. Jirků

Department of Fermentation Chemistry and Bioengineering
Institute of Chemical Technology, 166 28 Prague 6
Czechoslovakia

Production of hydrolytic enzymes is one of the key steps in any process involving biodegradation of natural biopolymers. The use of immobilized microorganisms as an enzyme source generally eliminates the high costs of enzyme production. In this context, studies on the production of β-glucanase system by immobilized <u>A. tenuissima</u> mycelium are reported.

MATERIALS AND METHODS

Microorganism and Cultivation

<u>Alternaria tenuissima</u> was isolated at the author's Department. The strain was grown in the medium described by Mandels and Weber (1969). The inoculum preparation was carried out at 24°C on a rotary shaker operating at 90 rev/min. Fermentor cultures were inoculated with 50-h-old mycelium at 20% of the fermentor volume.

Activity of β-1,3-glucanases

Samples of the medium filtrate were tested for laminarinase against laminarin (Calbiochem) and for β-1,3-glucanase using its affinity for yeast glucan extracted and purified from stationary cells of <u>Saccharomyces cerevisiae</u> according to Manners et al. (1973). One enzyme unit (U) is defined as the amount of enzyme which released one μmole of reducing sugar per min and ml of sample.

Gel Filtration Chromatography

Low-molecular components were removed from cell-free filtrates of the culture by rotary dialysis and the filtrate was concentrated by ultrafiltration. Proteins were precipitated with $(NH_4)_2SO_4$ (80% saturation) and the precipitate was dissolved and desalted on a Sephadex G-25 column. The protein fraction was concentrated in Aquacid III to a volume of about 1.5 ml which was used as a sample for molecular sieving on a Sephadex G-75 Superfine column (9 x 600 mm).

Support

Sorfix was prepared by polymerization of 2.6-dimethylphenol and conversion of the polymer into powdered porous form according to Kubánek et al. (1978). The specific surface area of the resulting product was 550 m²/g and the particle size was 0.2-1.2 mm; the pore volume was 0.459ml/g and the porosity was regular. The maximum distance of pores is 6.3 nm. Sorfix is noted for a high sorption capacity towards aldehydes which makes it possible to prepare "active" support with variable level of aldehyde groups.

Activation of Sorfix

The washed polymer (15 g) was suspended in 100 ml of 7% (v/v) glutaraldehyde. After stirring the polymer for 24 h it was transferred to a column and washed with water until the reaction with 2,4-dinitrophenyl-hydrazine, indicating free glutaraldehyde, was negative. Afterwards the polymer was washed with isotonic saline.

Immobilization Procedure

The mycelium was washed with isotonic saline, resuspended in mineral salts medium with glucose (1% w/v) and mixed with a suspension of activated support. The mixture was stirred at 20°C for 20 h and free mycelial forms were separated. The amount of bound dry weight per gram of carrier (the efficiency of immobilization) was determined on the basis of nitrogen quantity by Kjeldahl's method.

Immobilized Mycelium Column

The continuous-flow reactor consisted of a reservoir, a pump, and a jacketed tubular column (2.5 cm inner diameter; 50 cm length); temperature was controlled by circulating water. A packed column (4000 mg of bound dry weight) was recirculated at 30°C by 100 ml of mineral salts medium at different flow rates.

RESULTS AND DISCUSSION

Proteins of cell-free filtrate of a culture obtained after a 3-d cultivation in the presence of glucose were separated on a Sephadex column (Figure 1). The procedure yielded three active fractions of β-1,3-glucan-ase (I, II, III), the elution volumes of which correspond to those of the protein fractions A, B and C. The glucanase activity of fractions A and C cannot be demonstrated when yeast glucan is used in the test instead of laminarin. In addition, extracellular production of the β-1,3-glucanase system is constitutive and insensitive to catabolite repression. Laminarin and yeast glucan are degraded by neutral β-1,3-glucanase, the optimum temperature being 30°C.

Immobilized <u>Alternaria tenuissima</u> cells were prepared by a procedure ensuring the maximum proportion of the bound cell fraction. Preliminary experiments revealed that the efficiency of immobilization is markedly affected by the manner of carrier activation (reaction time and glutaralde-hyde concentration). The age of mycelium has no significant effect.

The time course of the glucanase system production is shown in Figure 2. It is evident that not only the total amounts but also the pattern of laminarinase I and III production is not influenced by flow rate. The oscillation in β-1,3-glucanase II production is enhanced with increasing flow rate. No deterioration of the structure of the column, morphological changes of immobilized mycelium, or changes of flow rate out of the column occurred during continuous 200 h use.

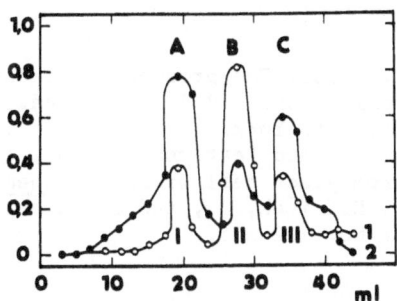

Fig. 1. Gel filtration chromatography of proteins of the filtrate of A. tenuissima culture; 1: protein content (absorbancy at 280 nm); 2: activity of β-1,3-glucanase (U/ml x 0.01); A, B, C protein fractions; I, II, III fractions of β-1,3-glucanase.

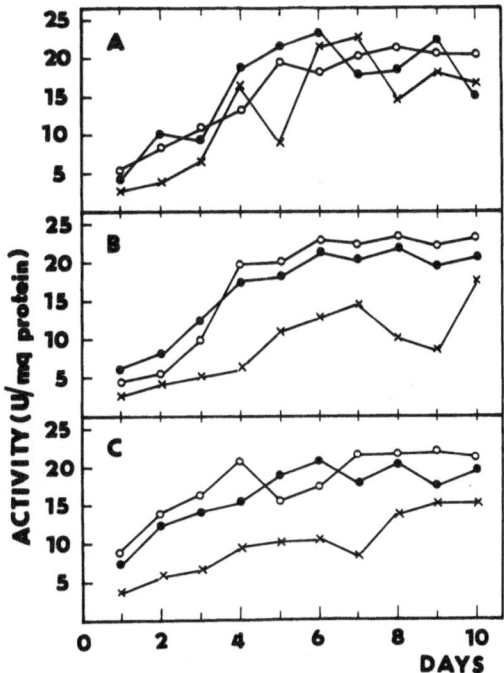

Fig. 2. Elution of A. tenuissima β-1,3-glucanase system from immobilized mycelium column. The column was run at a flow rate of 2.2 (A), 1.5 (B), and 0.7 ml/min (C); laminarinase I (o); laminarinase II (●); glucanase with affinity for yeast glucan (x).

From the experiments the following conclusions can be drawn:

a) A. tenuissima mycelium has the capacity of attachment to the activated particles of SORFIX, developing a porous network;
b) this attachment is based on covalent binding of mycelial cells via glutaraldehyde;
c) the binding of mycelial cells is determined by the amount of bound glutaraldehyde;
d) A. tenuissima mycelium exposed to the immobilization procedure retains the capacity of extracellular production of the β-1,3-glucanase system.

REFERENCES

Mandels, M., and Weber, J., 1969, The production of cellulases in "Cellulases and Their Applications", R.F. Gould, ed., American Chemical Society Publications, New York.

Manners, D. J., Masson, J. L., and Petterson, J. C., 1973, The structure of a β-1,3-D-glucan from yeast cell walls, Biochem. J., 135:19.

Kubánek, V., Veruovič, B., Králíček, J., and Cimburek, I., 1978, Czechoslovak Patent No. 202 215.

EXPORT OF ENZYMES INTO CULTURE MEDIUM BY

YEASTS OF SACCHAROMYCES GENUS

A. B. Tsiomenko, V. V. Lupashin, and I. S. Kulaev

Institute of Biochemistry and Physiology of Microorganisms
USSR Academy of Sciences
Pushchino, 142292, USSR

Export of invertase, acid phosphatase, β-glucanase, and some other proteins by intact yeast cells cultivated in liquid nutrient medium has been studied.

We consider the export as a terminal stage of a secretory process that includes passing of proteins secreted through the cell envelope followed by their exit into the cultural medium.

When cells were cultivated in YNB medium with ammonium sulfate as a nitrogen source, no protein export was observed with all strains studied. On the contrary, a certain amount of enzymes was exported in 2% peptone medium (Table 1). The rate of secretion was maximal during the whole exponential stage of growth. The highest level of activity was observed in the stationary stage of growth for all the enzymes under study.

The total number of polypeptides in the culture medium was 8-20 bands according to electrophoretic analysis (Figure 1). The protein export was inhibited by actinomycin D, cycloheximide, 2-deoxy-D-glucose and also by concanavalin A (Con A). In the presence of Con A (400 μg/ml), the enzymes were accumulated entirely inside the cells but not in the periplasm or the cell wall (Figure 2). At the same time the lectin had no influence upon cell growth and expression of invertase and acid phosphatase associated with the cell envelope. The activity accumulated inside Con A - treated cells was released into the culture medium on α-methylmannoside addition (Figure 3). The process was not affected by cycloheximide but was inhibited by sodium azide. The effect of Con A appears to be conditioned by its interaction with the plasma membrane of intact cells. Besides Con A, we detected peroxidase interaction with the plasmalemma of intact cells and alkaline phosphatase internment under physiological conditions.

It is known that the yeast cell wall permeability permits penetration of molecules with M_r of about 3-5 kDa. Yet the M_r of proteins exported as well as of exogenous proteins that penetrate through the cell envelope are much higher than the limit indicated. It was assumed that during yeast growth in peptone medium the permeability of the cell wall increased. To check the assumption, we compared the cell wall permeability of yeasts grown in peptone medium and exporting enzymes (+ ex) with the permeability of yeasts grown in ammonium sulfate medium and possessing no ability of export (- ex). It turned out that the cell walls of + ex and - ex cells

Table 1. Enzyme Activities in the Yeast Culture Medium.

| | Activity, mU/ml | | |
Strain	acid phosphatase	invertase	glucanase
1. Sacch.cerevisiae Y-350	80-90	500-1000	1-5
2. Sacch.cerevisiae D-15	120-150	2000	2-7
3. Sacch.cerevisiae Y-332	90	130-480	3
4. Sacch.cerevisiae D-14	100	200	-
5. Sacch.cerevisiae D-19	120	370	3-4
6. Sacch.cerevisiae CCY-21-4-13	100	280	2
7. Sacch.cerevisiae 1G-P-188	30-40	700-900	5
8. Sacch.carlsbergensis D-2	10-15	100-150	100
9. Sacch.carlsbergensis CCY-48-76	100	200-500	2
10. Sacch.terrestris Y-869	100	1000	-

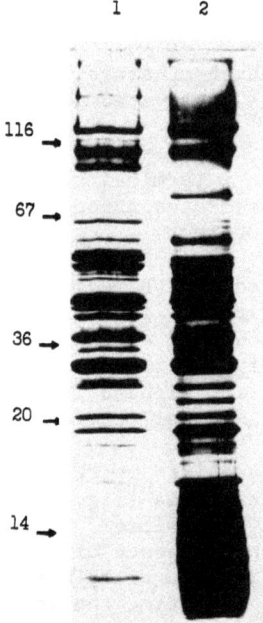

Fig. 1. Protein exported into the culture medium of S.cerevisiae Y-350.
SDS-PAAG gradient gel 7.5-15%. Numbers on the left refer to
molecular size markers (M_r x 10^3). Lane 1 - culture medium after
3 h protoplast cultivation; lane 2 - culture medium after 12 h of
intact cell growth.

had the same permeability for molecules with a M_r limit of 2-3 kDa (Figure
4). This was found in experiments with a number of dextrans and poly-
ethylene glycols.

It shall be recalled that the M_r of invertase and acid phosphatase is
about 250 kDa, and the M_r of the internalized alkaline phosphatase is 60
kDa. To explain the discrepancy between the values of the cell wall
permeability measured experimentally and the actually observed export and
import of macromolecules, we put forward the following proposal:

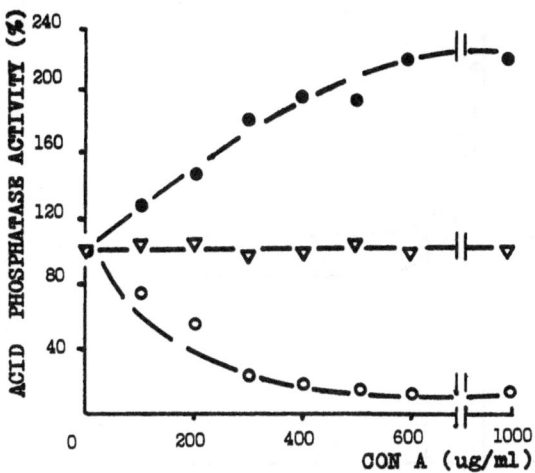

Fig. 2. Effect of Con A concentrations on the export of acid phosphatase
by intact S.cerevisiae Y-350 cells. ○ - exported activity; ▽ -
activity in the cell envelope; ● - intracellular activity.

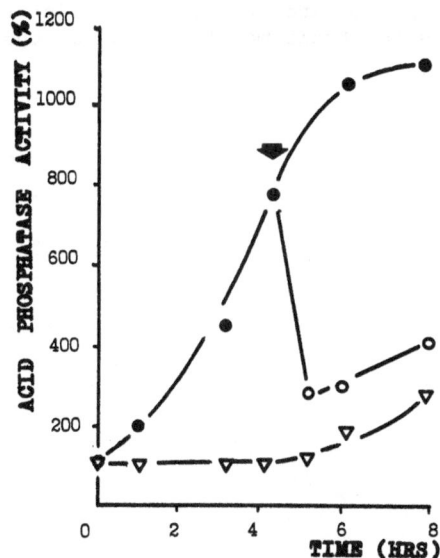

Fig. 3. Release of accumulated acid phosphatase from Con A - treated (400
ug/ml) cells. The arrow indicates α-methylmannoside (0.5 M)
addition. ● - activity inside Con A - treated cells; ○ - after
α-methylmannoside addition; ▽ - control cells.

The yeast cell surface is heterogenous. The cell wall represents the
greatest part but still only a part of the cell surface, and it fulfills a
protective function. The lesser part of the cell surface represents pores
or channels with abnormally high permeability as compared with the main
surface. These pores serve as communicational channels for active traffic
between the cell interior and external medium.

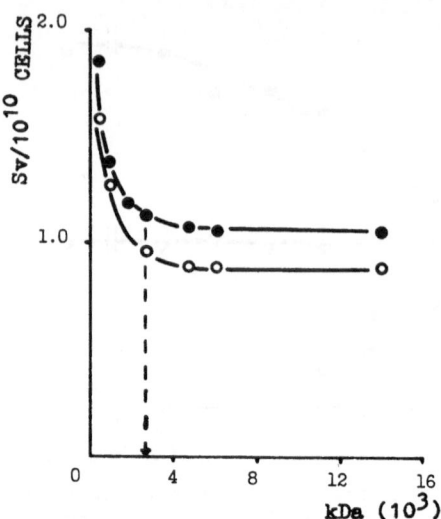

Fig. 4. Cell wall permeability of S.cerevisiae Y-350 grown in + ex (●) and
- ex (○) conditions. Sv - periplasmic space volume.

    In other words, the cell surface possesses some zones for such
important cell functions as exo- and endocytose. Owing to the active
nature of these two processes, their manifestation depends on the metabolic
state of the cell.

INDEX

Staphylokinase (continued)
  yield, per bacterial cell, 34
Starch
  as carbon source, fungi, 143-145
  induction of amylase, 3-4
Stationary phase, derepression of
      enzyme synthesis, 8-10
  (see also Cell cycle)
Streptokinase, 29
Streptomyces spp
  immobilization, 46
  inhibition of sporulation, 93, 94
  protease inhibitor, amino acid
      content, 96
Streptomyces granaticolor
  β-D-glucosidase activity,
      localization, 181-185
Streptomyces hygroscopicus, amylase,
      induction, 4
Streptomyces limosus, amylase,
      synthesis, 3-4
Streptomyces rectus, proteinase, 66
Streptomyces venezuelae, 181
Subtilisin-Amylosachariticus
  mutation distances, 63
  phylogeny, 66
Subtilisin-BPN, 59
  homologies with thermitase, 67
  phylogeny, 62-63, 66
  prediction, α- and β-structure,
      68-69
Subtilisin-Carlsberg
  prediction, α- and β-structure,
      68-69
Subtilisin-DY
  activity, 100
    determination, 101
  chromatography on bacitracin -
      cellulose, 101
  mutation distances, 63
  purification by bacitracin -
      cellulose, 99
Subtilisins
  action of protease inhibitors, 94
  characterization, 59
  thermitase, and proteinase K,
      sequences compared, 64-65
Subtilisin-type proteinases
  active site histidine residue, 61
  cysteine-containing, 66
  phylogenetic tree, 66
  (see also Thermitase)
  table of mutation distances, 63-64
Sucrose, levansucrose synthesis, 6

Temperature changes, effect on
      proteinase formation, 80
Thermitase, 59-70
  amino acid sequence, 60
  cysteinyl residue, 59, 61
  homologies

Thermitase (continued)
  homologies (continued)
    B. cereus proteinase, 67
    subtilisin BPN, 67
  mutation distances, serial
      proteases, 63-64
  phylogeny, 66
  prediction, α- and β-structure,
      68-69
  purification, flow diagram, 61
  spatial configuration, 66
  subtilisin, and proteinase K
      sequences compared, 64-65
  thermostability, 69-70
Thermactinomyces vulgaris, source of
      thermitase, 59
Thermomyces lanuginosus, α-amylase
      production, 143-145
Thermophilic bacteria, see B.
      stearothermophilus
Transcription, inhibitors, and enzyme
      production, 89, 92
Transformation/cotransformation, in
      A. niger, 52-53
Trichoderma reesei
  enzymes, 191
  β-glucosidase, 173
  multienzyme cellulases, 154
Trichoderma reesei QM6a, 157
Trichoderma reesei RUT C-30
  cellulase secretion, 157-171
    catabolite repression, 157-158,
      171
    endoglucanase, 160, 165-171
    fractionation, 159
    Golgi apparatus, 162, 164-167
    growth, comparison with wild
      type, 161
    immunoprecipitation, radio-
      labelling, 169-170
    induction, endoplasmic reticulum,
      162-164
    summary and discussion, 170-171
Trichoderma viride, see T. reesei
Tritirachium album, proteinase K, 59
Tryptone Soya Broth medium, 15

Umb-G, fluorogenic substrates, 187
Umbelliferyl glycosides, fluorogenic
      detection, enzymic activity,
      187-191

Vector plasmids, construction,
      122-125
Vegetables, pectinolytic enzyme
      action, 48

Xylan, cellulolysis, 187
Xylan, differential affinity of
      endoglycanases, 189